The Employer Safety Guidebook

to

Zero Employee Injury

Third Edition
March 2011

by

Emmitt J. Nelson, ME, PE, NAC, ZIC

The Employer Safety Guide Book
to Zero Employee Injury
Third Edition

The Employer Safety Guidebook to Zero Employee Injury
Third Edition
March 2011

Published by:

Nelson Consulting, Inc.
10031 Briar Drive
Houston, Texas 77042

Printed in the United States of America

ISBN 978-0-9791685-4-3

The Employer Safety Guide Book
to Zero Employee Injury
Third Edition

APPRECIATION - 2003 FIRST EDITION

Writing a book is a long term project. Parts of this book have been in process for at least 20 years. The work in its present form has not made it to press before now for many reasons, but I feel it can wait no longer. It is clear that it has been greatly enhanced by the material made available since its inception and can clearly be a much more powerful contribution to the business world now than had it been published in a prior year.

I owe a large debt of gratitude to many of my industrial colleagues, both in Shell Oil Company and others who, sometimes unwittingly, but powerfully impacted the content of this effort through their many encouragements as we have walked this path of bringing the Zero Injury message to the construction industry in particular and along with this to companies at large no matter their business enterprise.

My thanks also goes to Dr. Richard L. Tucker, currently occupant, Joe C. Walter Chair in Engineering, The University of Texas, Austin, for his in-depth critique and recommendations on a host of matters pertaining to the content and presentation of this work.

Also my deepest gratitude goes to my wife Ginny for her unfailing encouragement and to our Granddaughter Christina Ashie, an Honors English student, for her untiring editing. If flaws remain in this book they are caused by my failings and not these who have selflessly made their earnest contribution to its success.

Four construction industry colleagues volunteered to read and comment on the manuscript. I wish to thank them all. Excerpts from their comments may be see on the back cover.

Emmitt J. Nelson

APPRECIATION - 2011

A lot of safety water has "flowed under the *Zero Injury Bridge"* since the First Edition of <u>The Employer Safety Guidebook to Zero Employee Injury</u> was published in 2003, now known as "The Green Book." Only the most astute visionaries would have dreamed in 2003 that the CII members would be boasting a TRIR of only 0.63 in the year 2009 versus the national average of 4.3. Incidentally the 0.63 applies to 1,117,000 workers or 2.2 Billion hours worked, not a small sample. I will speak of this record in the updated text.

There is a deep sense of gratitude in my heart and I want to express my appreciation to all who have purchased copies of the First Edition for you gave me inspiration to keep writing on the subject of the Zero Injury notion and the amazing safety performances registered literally around the world under the banner of Zero Injury. I also want to thank those who have placed their faith in me and became one of my clients during these past 20 years of safety consulting.

And now in 2011 I also want to thank two friends, Bennett Ghormley Sr., and Tracy A. Hanes, who asked me to join them in founding the "Zero Injury Institute" (ZII). ZII's purpose is to continue to encourage the use of the Construction Industry Institute's zero injury research and to specifically perpetuate the practical application knowledge gained from the practitioners of The Zero Injury Safety Leadership Concept.

Bennett and Tracy approached me in early 2009 with the proposition to join them in forming the Zero Injury Institute. With their encouragement I joined in the effort as "ZII Associate and Subject Matter Expert." Since then we have developed a battery of products to offer clients including Zero Injury Seminars, Workshops, Safety Culture Assessments, Employee Safety Perception Surveys and Zero Injury presentations of various lengths often tailored to the specific client needs.

Also for the purpose of maintaining critical sustainability of a zero injury culture ZII offers on-line safety knowledge certifications with eLearning in the Zero Injury Safety Leadership Concept. This medium is in modular format of one hour each and is available for Crafts, Foremen, General Foremen, Superintendents, Safety Specialists and Project/ Construction Managers. The module content builds on one another in such a way that each higher level position must first take the modules for the lower level positions in order to advance to the selected level of certification. Each module ends with critical information on the position specific role elements of each individual as the safety culture evolves into one where at-risk behaviors decrease and incident/injury becomes extremely rare events.

Joining ZII in alliances to provide this training is The Association of Union Contractors, (TAUC) of Arlington, VA, The Houston Area Safety Council (HASC), Deer Park, Texas. In addition Nelson Consulting, Inc. has agreed to provide the Safety Council of Texas City with these training services as well.
See www.zeroinjuryinstitute.com.

Lastly, I wish to thank my family for their encouragement and support as I continue this journey of being a zero injury safety zealot. Twenty years ago I would have never dreamed in my closing years I would become something I would have never dreamed; an enthusiastic salesman of safety!

Thank you one and all.

Emmitt J. Nelson,
March 2011
Houston, TX

ABOUT THE AUTHOR

Emmitt J. Nelson, President of Nelson Consulting, Inc. and Associate to the Zero Injury Institute began a safety consulting business after retiring from Shell Oil Company in 1989. He has authored and co-authored numerous articles on the Zero Injury Safety Leadership Concept for trade magazines and also authored two books on construction safety management featuring the Zero Injury concepts this book's third edition being one of the two. The Employer Safety Guidebook to Zero Employee Injury and The Pathway to a Zero Injury Safety Culture.

During the last five years of his career with Shell Oil Company Emmitt served as Corporate Manger of Construction Relations and Engineering Standards. In this capacity he led a corporate emphasis on bringing contractor safety performance into line with Shell's own employee safety performance. To accomplish this he led in the creation of a corporate-wide contractor safety strategy and presenting it to both Shell's and contractor management leadership.

Immediately upon retirement from Shell, contractors interested in reducing injury to their employees enlisted Emmitt as a consultant. In 1992 he founded Nelson Consulting, Incorporated, a Texas S Corporation specializing in teaching the Zero Injury Safety Leadership Concept. The business purpose is to assists clients in the installation of the safety management processes required to create a working culture where injury is a rare event.

The best of his clients have achieved over 4,500,000 work hours with Zero BLS/OSHA Recordable Injuries. At least two construction contractor clients have completed over 2,000,000 work hours with Zero Recordable injuries in the USA, with one achieving over 2,500,000 hours with "Zero Recordable" injuries and the other achieving over 4,500,000

Emmitt is a B.S. graduate in Mechanical Engineering from Texas A&M University, and a Registered Professional Engineer in the State of Texas.

In 1999 Emmitt was honored with the "Construction Industry Safety Excellence" award by the Business Roundtable.

In 2000 he was honored for his construction industry wide contribution to safety excellence with election to membership in The National Academy of Construction.*

In 2003 he was honored by being named an "Outstanding Instructor of the Year" by The Construction Industry Institute.**

While employed by Shell Oil Company, he was Shell's representative to The Business Roundtable*** Construction Committee where he served as Committee Chairman for two years.

Additionally he served as Co-Chairman of The Center for Construction Education, at Texas A&M University.

He also served as Chairman of the Construction Industry Institute Zero Accidents Research Taskforce from 1990 to 1994. This Taskforce performed the initial CII research into the Zero Injury phenomenon uncovering the basic safety management strategies used by those employers who are successful in working millions of hours with zero injury.

In 2009 Emmitt became associated with The Zero Injury Institute**** as Co-founder and Subject Matter Expert, (SME).

* The National Academy of Construction (NAC) is an independent professional organization whose new members are elected to NAC through nomination by and balloting of the current members. NAC honors construction industry

leaders whose professional career over a period of years has demonstrated an outstanding contribution to the improved effectiveness of the engineering and construction industry in the USA. Emmitt was elected in recognition of his leadership in the research into how zero employee injury is achieved and for his "promotion and facilitation of a fundamental change in the Construction Industry performance in worker safety."

** The Construction Industry Institute (CII) conducts research for the purpose of making the American Construction industry more effective in the world marketplace. The members of CII are owners that use, the contractors and construction service organizations that provide America with construction services. As of 2009 there are ~105 members, about one half owners and one half contractors with a few service organizations joined in. The institute is affiliated with The University of Texas in Austin, Texas.

CII members provide research team members to task forces that perform batteries of research on a broad range of construction topics with the purpose of improving the American construction project process from conception to completion and beyond. Members are able to benchmark their project performance against the best in the construction business thereby are able to gain prominence in the industry.

*** The Business Roundtable (BRT) is made of about ~200 of the largest businesses in the USA. The CEO's of member companies make up the BRT. Senior "Member Company" executives make up the several committees that function under the umbrella of the BRT. The Construction Committee (now disbanded) was one such committee that engaged in creating a more efficient and effective construction industry in the USA. This committee functioned from 1972 to 2000, a period of 28 years. During this time the BRT Construction Committee completed a number of significant industry changing accomplishments. One of these was the CICE (Construction Industry Cost Effectiveness Project,) another was the CISE (Construction Industry Safety Excellence) awards. The latter was implemented while Emmitt was Chairman of this committee.

**** The Zero Injury Institute (ZII) was founded in 2009. ZII's purpose is to provide knowledge training in the Zero Injury Safety Leadership Concept. ZII does this through on-line zero injury safety training modules designed by Emmitt to provide position specific knowledge about the zero injury concept in a way that allows the student successfully completing the training to function effectively in a culture of safety where zero injury is the actual daily outcome. When all employees in all locations a where company may perform work all can be equally informed.   It is then company wide safety outcomes can steadily improve until all injuries are being prevented. The eLearning modules are available for Craft, Foremen, Superintendents, Project Managers and Safety Specialists. See www.zeroinjuryinstitute.com

Web sites for reference are as follows.
www.zeroinjuryinstitute.com   www.construction-institute.org   www.nelsonconsulting.com
www.naocon.org

PREFACE - First Edition – 2002
by Dr. Richard L. Tucker,
Joe C. Walter, Jr. Chair in Engineering,
The University of Texas, Austin

I am honored to provide my perspective on "The Employer Safety Guidebook to Zero Employee Injury." Perhaps a bit of history will be useful in providing the context of this important work.

While I served as the Founding Director of the Construction Industry Institute, a research organization involving many of our nation's largest owners, contractors and universities, the concept of Zero Accidents came into being. CII was itself a pioneer effort and, in its early days, its industry representatives met regularly to identify important research areas. At one of these meetings, Keith Price, Executive Vice-President of Morrison-Knudson, stated "We need to address construction safety. We must find ways to reduce injuries on our job sites." Thus, CII began its safety research studies in the mid-1980's. Expectations were modest, since the industry's safety statistics had not improved for many years.

Emmitt and I were not well acquainted at that time, and he was not involved in the initial CII safety research. As representative for Shell Oil Company, he was elected as Chairman of the Business Roundtable's Construction Committee, an organization known for its earlier "Construction Industry Cost Effectiveness Project." With the support of John Bookout, CEO of Shell Oil Company, the Business Roundtable Construction Committee, under the Directorship of Richard F. Kibben, launched a major safety initiative. That initiative, built around safety excellence awards and recognition, for owners and contractors, significantly raised the visibility of safety in the owner community. As this effort progressed, Emmitt's passion for safety improvement became increasingly obvious. He retired from Shell Oil Company after completing his term as chairman of the Construction Committee, but continued his drive for safety improvements as a safety management consultant.

Shortly after Emmitt's retirement, the CII began considering additional safety research. It even set goals for safety improvements for its companies to achieve by the year 2000. (The goals were modest; only a 25% improvement.) At one of the CII Board of Advisors meetings, a member stated "We must find ways to eliminate all accidents on our job sites. Only zero accidents is acceptable!" Thus, a name and charge were born for a new research effort. I then contacted Jim Braus, a Shell executive, and asked if the company might re-hire Emmitt, as a consultant, to lead the Zero Accidents Research Team. Shell readily agreed.

As they say, "the rest is history." Emmitt's passion was shared by the other team members. They brainstormed and formulated ways to generate a "zero accidents" mentality throughout the industry. I felt that some of their ideas were impractical and cautioned against their implementation. They didn't take my advice, and they were right. The CII videotape, "One Too Many" and its associated Zero Injury publications rapidly became CII's most widely adopted products. Many CII member company safety records improved dramatically, as illustrated in this book.

Emmitt's devotion to Zero Injury in employee safety continues. He is one of our industry's most noted safety consultants and has been recognized by election to the National Academy of Construction. This book will not be the culmination of his efforts, because he will continue to be active. However, it provides a seminal contribution to our drive for safety improvements.

It is possible to have zero accidents, and this Guidebook tells us how!   Richard L. Tucker, September 19, 2002

INTRODUCTION BY AUTHOR - 2011

Some may wonder about the Second Edition of _The Employer Safety Guidebook to Zero Employee_ Injury and what happened to it. The Second edition was a Paper Back partial (not all material was included) edition published by the Zero Injury Institute via Amazon Digital Press. It is no longer available

Yes, this is the third edition of The Employer Safety Guidebook to Zero Employee Injury. In the construction industry it is called the Green Book by those who have purchased it. I like the reference to Green because I feel the name recognizes an allegory; the Green way of leading safety. It tells how to avoid the tragic waste of the worst kind; the waste of human dignity through workplace injury and yes, even death.

This then is the Third Edition which will include selected revisions to the narrative along with updated safety stats where appropriate. This book was the first of two books I have published on the Zero Injury Safety Leadership Concept the other being _The Pathway to a Zero Injury Safety Culture_ published in 2005.

It was in 1987 that The Business Roundtable Construction Committee gave the first Construction Industry Safety Excellence awards. It was noted that two of the awardees, an owner and a contractor, had performed their work without Lost Workday cases for two years and four years respectively. These amazing achievements for that time were soon labeled the "Zero Injury" phenomenon borrowing from author Philip Crosby's "Zero Defect" approach to Quality.

The significance of this zero lost workday case performance was most startling when compared to the OSHA National Averages for worker injury in construction in the year 1986 of 6.8 lost workday cases per 100 workers. Comparing the

OSHA national average to these awardees that worked a combined 6.8 million work-hours with no Lost Workday Cases, the two companies avoided 231 serious injuries to their workers. Wow!

When these achievements were announced by the Business Roundtable many asked, "How do they do that?"

Soon after, the Construction Industry Institute (CII), located in Austin, Texas, and led by Dr. Richard L. Tucker, agreed to commission a research Taskforce to look into the question of: "How do some contractors and owners achieve millions of hours worked in construction without serious injury?"

The research Task Force Chaired by Emmitt was named the "Zero Accident Taskforce." The work was completed in 1993. In 1999 CII decided to follow-up with an investigation of these "Zero Recordable" achievements; an even more amazing safety performance than "Zero Lost Workday Cases" investigated in 1993.

Additional research was accomplished and reported out in 2001 and 2002. These data verified and expanded on the 1993 results yielding additional quantifiable data that reflects significant safety performance improvements in companies that have utilized and improved upon the original Zero Injury techniques. Many of these have achieved Zero Recordables for extended numbers of work hours. The CII "Making Zero Injury a Reality" Task Force restated the injury original five critical safety elements, increasing the number from five techniques to nine.

The third edition of the Green Book describes the successful application of the CII research derived Zero Employee Injury safety management techniques and introduces a employee involved safety management process that helps achieve the most critical aspects of creating a working culture and process where employee injury is rare.

The Employer Safety Guide Book
to Zero Employee Injury
Third Edition

TABLE OF CONTENTS

PART 1

AN IN-DEPTH EXPLANATION OF THE ZERO INJURY CONCEPT

CHAPTER 1

THE SEARCH FOR ZERO INJURY

Safety Excellence Re-Defined

In the early 2000's _Zero Injury_ as Safety Watchwords had reached worldwide use. All around the globe employers were adopting the concept of striving for Zero Injury. The problem then and now is many have not also adopted the proven research of the Construction Industry Institute and adapted it to their work setting.

While many have found the zero injury process, there are still many more for whom the Search for Zero Injury continues; for them the questions are the same as they were seven years ago.

Where is the trailhead; where do we start; do you have a map?

The answer to these questions is still right here in this Guide Book; the "trailhead" begins here; you have the map in your hands!

Take a look!

In the past 17 years the words "Zero Injury" have come to mean that long sought superior competitive safety performance that enlightened executives wish to lead their organizations to achieve.

In much of American industry, the decade of the 1990's was spent in pursuit of safety excellence. The decade of 2000-09 was one where "those that were still seeking" pursued those "who had found" the keys to a Zero Injury workplace.

A Shadow Looms in 2011

As this third edition of the Green Book is being created there looms a bank of ominous clouds casting an unfortunate shadow over the use of the Zero Injury research of the Construction Industry Institute (CII).

This shadow is created by leaders who have heard about zero injury and are demanding a zero injury outcome while ignoring the proper application of the CII research which has been proven many times over to produce a zero injury outcome when applied properly.

Many people in business and industry struggle with defining their personal answer to the following two questions.

What is safety excellence? Can it be defined?

A good definition of safety excellence is this.

> *"A company or group of employees*
> *have reached safety excellence when*
> *all 'at-risk behavior' has been eliminated."*

In contrast, however, repeatedly the definition of "safety excellence" commonly found in business and industry is as varied as there are companies. Too often leaders accept mediocrity (typically un-intentioned) as their definition of safety excellence.

All too many say to themselves "Not to worry, we are doing everything we can. After all, we are better than the OSHA/BLS national average."

Or those above the national average might say, "If we can just get our performance in line with the national average we will be fine."

The redefinition of safety Excellence came to us as a product of the decade of 2000-2009. This definition of safety excellence is now "Zero BLS/OSHA recordable injuries while working a million hours or more." The resulting Total Recordable Incident Rate is 0.20 or less. This compares to a BLS/OSHA national average for the year 2009 of 4.3.

Since an "OSHA/BLS average" is the arithmetically combined and normalized safety statistics of all in an industry group there are those below average and there are those above average. It is commonly found that in any industry group there are at least a few companies, some with as many of 2000+ employees, who experienced no recordable injuries for a 12 month period and then continued the record run for another 2 months. This occurred from October 2003 to December 2004. This was accomplished by S&B Engineers and Constructors of Houston, Texas.

Despite this type record for own record comparison, it seems the approach most safety executives take in measuring their company's safety performance is to compare their data with the national average. As they do this, more often than not, they do not even stop to realize that there are those in their industrial group without an injury. These perform at Zero Injury!

Why settle for the average? Why do we not measure our performance against the best: those with no injuries?
Here, the key questions should be, "Who are they, and how

do they accomplish such performance?"

This Green Guide Book Reveals

One of the things this Guide Book does is to identify who these successful companies are in one industry, the construction industry, and goes on to reveal exactly how they are attaining such lofty safety performance levels.

Hopefully this information will cause those not in the construction industry to ask the following question.

"Why should I read what those in construction are doing when I am in a different industry?"

The answer is in simple terms: employee safety is a people leadership and management challenge. While the hazards presented in various industries are different, the people are the same. Copy what these in construction are doing in leading and managing safety with their people and you, too, can achieve Zero Injury.

Safety: A Psychological Game

While I do not view safety as a "game" the phrase application applies. No matter the industry, above all else safety in the workplace is a mental game, a thinking game or "mental safety" if you prefer.

How many times have you heard after an injury, a statement by the injured explaining what happened similar to the following? "I just wasn't thinking."

How many times have you heard leaders say, "If we could just get everyone to be on the alert and mentally engaged in their own safety all the time we would be OK."

Such a view of "mental safety" gives significant insight into the principal challenge faced in creating a working culture where "risk taking" is a rare event. In the realm of "mental safety," there is an additional aspect to be addressed. In addition to the first, the "thinking element" there is also the second, the "knowing element."

Here, I am simply saying that, in practicing safety an employee cannot think of something they do not know; thus, as far as safety training is concerned, employees are unable to think of at-risk behaviors to avoid they have not been trained to avoid. Of all the zero injury workshops conducted over 17 years I can truthfully say that the attending employees, leaders or workers have always chosen "zero injury safety training" among the top three priorities to be used in converting a traditional safety culture to one where incident/injury are very rare events.

To create the possibility of Zero Injury occurring in your culture there are two primary challenges remaining after the safety training is complete. These challenges are 'motivational." While training is of paramount importance and is an absolute requirement for a sound zero injury safety culture, motivating employees to (1) "think safe" and to (2) "follow that safe thinking" with "safe behavior" is the bottom line challenge all, management and worker alike, face in creating a workplace free of injury. In order to meet this challenge successfully we must believe we can! Too many are simply engaged in disbelief thus forever voiding the "possibility" of achieving zero injury.

Some disbelieve because they remain focused on the sheer statistical "improbability" of zero injury occurring because people are fallible. Yet the statistical "improbability" can never remove the statistical "possibility." It is because of this focus on "possibility" that we see so many contractors and owners realizing these 1,000,000 to 5,000,000 strings of hours worked without a BLS/OSHA recordable.

As an example, while speaking with Greg Rosier, Safety Director for Cherne Contracting, a Minneapolis, MN construction firm that had just exceeded 2,500,000 work-hours (August 2001) with Zero Recordable injuries, Greg made the following simple observation.

> "After one has implemented all key zero injury safety management techniques, success in achieving a zero injury workplace is 99.9% between the ears. Believing you can is critical. Too many people do not believe they can achieve such lofty dreams," he said. Continuing he observed "The simplicity of my point was illustrated to me in the first grade by the story about the little Steam Engine that puffed up the hill with the monologue 'I think I can, I think I can, I think I can, <u>I know I can</u>.'"

He added the following:

> "If people insist they cannot work injury free they will not."

It's a Culture Change

How does an organization go about changing the work culture to one where there is little to no "at-risk" behavior?

In a historic work culture, while injury is not wanted and all people work to prevent injury, nonetheless, when an injury occurs, the injury event all too often is treated as though it were inevitable! "Injuries do occur, after all" becomes the thought in everyone's mind. Please allow me to advance an allied observation that is essential to a Zero Injury culture.
<u>*The fact that injuries do occur, does not*</u>
<u>*mean that injuries must occur.*</u>

In analyzing for the root cause in "most all incidents/injuries" there is discovered a number of interventions that if accomplished would have prevented the event.

To illustrate this point allow me to give an example. A client found in investigating a fatality, an in-depth analysis of the events leading up to the death uncovered seven decision points in the chain of events as potential accident intervention points. Had any one of these seven preventive actions/inactions leading to the event been taken, the fatality would likely not have occurred.

You may have noticed two paragraphs above, I used the term "most all incidents/injuries." The use of these words recognizes that "on occasion" injuries may result from an "act of God," such as a sudden rainstorm or a lightning bolt.

In such cases one can be on the alert and remove themselves from areas of maximum exposure to minimize the chance of being injured.

By way of observation I see far too many people overusing the term "act of God." Some would say that if a machine fails leading to an injury such a failure was an "act of God." As an engineer, trained in equipment design, I disagree. For instance, if a machine is being used in such a way that exposure to injury or the possibility of failure exists it is incumbent on management to remove the possibility of an injury or to remove the possibility of a failure causing injury. Machine guarding is a common example of the former.

The key point I am making in the paragraphs above is the following observation:
> *The common enemy of a safe work culture is "in the mind" of company leaders, managers, foremen, union leaders and workers.*

One could say that in the minds of some people the possibility of working injury free is "locked behind the prison bars of tradition."

Based on past traditional experience the thinking is, "since injuries have always occurred then it is a given; injuries will always occur." This is both truth and fallacy! While it is true an injury will occur someday, those who achieve Zero Injury work with the simple commitment that "we will prevent all injury today."

It is a fallacy to think that an injury cannot be prevented.

As this fallacy becomes entrenched in the belief system, the inadvertent "mind-set" for many becomes: "I know injuries will occur. Beyond a nominal amount of effort to encourage the use of safe behavior on the part of the worker and complying with all legal requirements there is little 'else' to be done by those in charge."

With this mind-set, failure is inevitable, and accepted even before beginning the effort to address the subject of Zero Injury. This gives rise to make the above statement that even "the possibility of working injury free" is in fact locked up in our tradition. Just as locked up as if it were actually behind prison bars!

To have a Zero Injury workplace, this "injuries will occur" mind-set must be challenged and changed.

And here is the good news. Repeatedly, research has proven there is something "else" that can be done. It is found in embracing a new "mind-set" or "concept" on how to lead your employees to adopt your safety vision.

The Zero Injury Institute has now named the Zero Injury approach to safety as "The Zero Injury Safety Leadership Concept." When applied this research backed concept requires a company or project safety culture change. The traditional culture of accepting injury, as an unwanted byproduct of our business ventures must change to one where all employees, from the worker to the CEO, believe that injury can be prevented and all accept that challenge.

The Zero Injury Safety Leadership Concept was not invented by this author or by the Construction Industry Institute, the source of the research, but was invented by employers who dared to think the seemingly impossible.

A commitment to the Zero Injury Safety Leadership Concept, simply stated, is:

> *"Since no one wants an injury to occur and since any injury causes our employee to suffer and also harms our business plan we will do whatever it takes to prevent the injury of our employees."*

> *Simply put this "do whatever it takes" translates into leaders creating innovations that cause employees to avoid all at-risk behavior including the leaders as well.*

Such a stance is an example of how to define Safety Commitment.

CHAPTER 2

REDEFINING SAFETY COMMITMENT

The Question

All agree that, if an injury free workplace is to exist, management and worker COMMITMENT is a very essential and critical factor.

But what does the term "commitment" really mean when used to describe one's devotion to an injury free workplace? Has the term "commitment" become so all encompassing that it has no hard factual meaning? The term "committed" is often used in the following way as stated:

> *"Though I was committed before, back then I thought that all injuries could not be prevented, but I have changed. Now I am really committed. Not just a little, I mean a I am totally committed."*

It is commonly true and reasonable to state that all the people involved in a work process are "committed" to the thesis that "no one will want" injury to occur. But is this "no one will want" feeling everyone has, the only essential ingredient in defining our "commitment?"

On examination, the broad limitless range of meaning we find for the word "commitment" using words like underlined above; "really" and "totally" leads to an unsatisfactory answer. Just how "real is really" and how large is your "totally" compared to my "totally!" These words are mere "sponge" when one attempts to find definitiveness. Immediately, it is apparent that to define "commitment" we must go far beyond superficial spongy definitions.

The pivotal question then is, "Are we, the management, and the employees COMMITTED, "in fact," to a safe workplace, when we refuse to believe that injuries are preventable, or that we cannot work injury free for a larger number of hours? Think about it, working at a zero injury rate is the only way we can improve. All the progress made in the construction industry in the past 20 years as the TRIR has improved from 16.3 (1989) to 4.3 (2009) has been accomplished by working more hours free of injury. Yes "more hours free of injury" does equal "hours at zero!"

Embedded in the Zero Injury thesis is this fact:

We are not committed to an injury free workplace until we embrace the notion that all injury is preventable, that no injury is acceptable and that we will not sacrifice schedule, cost or production over an injury free workplace. Nor will we set injury goals, thereby inadvertently indicating that some number of injuries is expected and hence acceptable.

In recent years when asked to define safety commitment, most people would agree to the answer as follows,

"Safety commitment is in evidence when the workplace is free of at-risk behavior."

Where work places with very low numbers of at-risk behaviors are found, it is also found that management and employees alike are sold on the zero injury safety process.

Zero Incident/Injury Root Cause Analysis

It is currently common safety practice to perform 'root cause" analysis to determine the real cause of an incident/injury. The Root Cause process is a disciplined approach that minimizes the possibility to become distracted by emotion or what appears to be obvious.

Here I propose we do a Root Cause analysis of the long term (1,000,000 work hours) existence of an injury free safety culture.

Rather than "drilling down" into the actions/inactions of those on the worksite when an injury occurred lets reverse the process and "drill up." Drill up from the "presence of zero injury" to look for the base line actions by employees and leaders; to seek those safety techniques that are in place as the causative factors of the zero injury result. What would one find?

Looking above as we drill up, just a tad bit higher at the absence of injury one clearly sees the absence of "at-risk behavior."

Looking higher still above "at-risk behavior" what will we find as the basic cause of at-risk behavior?

Oh there it is, just a little higher up; I see it now; looks like "at-risk thinking" doesn't it?

Investigating further up the chain of cause and effect what will we find lurking just above "at-risk thinking?"

Two things we can see; 1. the "lack of safety knowledge" and 2. a risk taking attitude (lack of buy-in to working safe).

Analyzing these two in numerical order we find behind the root cause of 1. this "lack of safety knowledge" that something needed is missing. But this time this absence is caused by something we can do, as opposed to not do; the latter being avoiding at-risk behavior. It is by conducting appropriate and timely zero injury safety training for all employees. But some will say that is a massive task. The immediate question is: "What if I lose them?"

"Why I will be training employees who might only work a few days, some perhaps only one day. What can be cost effective about that?"

Dr. Metcalf has the following to say about such thinking:

> *"There is one thing worse than training employees and losing them and that is not training employees and keeping them."*

One simply cannot say who is going to stay how long on a job site. Every employer will have significant safety training to give new employees. The zero injury portion of safety training needs to be consistent in content. Right away we can see this missing training ingredient is correctable through the application of "designed safety training modules."

Of course these modules will need to be based on the foundational elements discovered by the CII research found to be essential zero injury safety culture building elements that experience has shown to be present in a zero injury safety culture.

Now on the occasions when a 2. "Risk taking attitude" is present it usually results from a tradition of "I/we have always done it that way" or "I/we need to take this risk in the interest of production or "I/we can get away with it this time" or "I/we can see this but I do not see how it can apply to our operations, after all we are different!"

There are others in this group of traditional risk takers but what is interesting none of them really see themselves as risk takers. As a consequence, individually or as a group, they are very difficult to convert to a zero injury mind set. The only way to convert them is "one at a time." With time and logic on the zero believer's side one need only be patient.

However, If the risk taker is your own employee you simply give them all the information and ask for buy-in. You explain the buy-in requested is required to retain your job. If an employee chooses known at-risk behavior they are simply "choosing not to work here." Why so harsh? If you think about it a zero injury culture and an employee with a consistent risk taking attitude cannot co-exist.

Remember we are considering root causes of a zero injury outcome here. Zero is found in the absence of risk taking. So if you truly want zero injury then you really have no choice; but I hasten to add you can be gracious and give them a choice.

Make their choice simple; say "here we require zero at risk behavior; if you engage in at-risk behavior you are choosing to not work here."

I call it "self-termination."

How Do We Change the Culture?

So the big question of all leaders is; how do I/we do that?

Leaders do it by establishing safety as a Value. It is no longer a priority. Priorities can change, Values do not.

Leaders do it by convincing all that nothing is more important than the absence of injury, not schedule, not production and not profit.

Leaders do it with education of all employees on the foundational elements of a zero injury culture. Inform all that nothing is more important than a safe outcome. That since no injury is acceptable then no at-risk behavior is acceptable, by leaders or workers.

Leaders do it by exhibiting a caring attitude toward workers, convincing them that nothing is more important than being injury free at the end of each and every day.

Leaders do it by applying the zero injury research basics; by allowing worker participation in the creation of a safe culture. Participation allows contribution; contribution used promotes safety culture ownership and ownership brings safety culture support.

What is the key leadership ingredient?

Which is tops in the crucial safety leadership techniques?

It is "Personal, individual by individual Commitment!"

CHAPTER 3

DEFINING ZERO INJURY

Defining The Zero Injury Concept

To begin this pursuit of a Zero Injury safety culture change, first allow me to define the term "Zero Injury."

Many people ask me if in using the term Zero Injury I mean that there can be absolutely "zero injury?"

In applying the basic Zero Injury Concept the answer is a clear yes. That is the concept. No injuries of any kind whatsoever! Zero!

At this point again the skeptics think such is impossible.

My simple reply is as follows:

"It is not impossible. You are already achieving Zero Injury. Unless an employer has an injury to one of their employees every day, some days are performed at the Zero Injury rate."

The question is not: "Is Zero Injury possible?" but rather is "How many hours can my group of employees work without an injury of any kind?"

Allow me to illustrate. Say you have a group of 100 employees. (This is a handy number to use because

BLS/OSHA measures injury frequency using a cohort of 100 employees.)

For this example let's take the Construction Industry in America. OSHA reported that in 1989 there were an average of 14.3 Recordable Injuries per 100 workers. This means hat in 200,000 work-hours an average of 14.3 injuries occurred.

Where do the 200,000 work-hours hours come from? Why use that standard to measure?

It is 100 employees working eight hours per day for 250 days. 100 x 8 x 250 = 200,000.

Where does the 250 days come from?

It is 365 day year less 104 weekend days less about 10 to 12 holidays. The result is approximately 250 days; give or take a day or two. So OSHA uses the 250.

Getting back to our 14.3 injuries in 200,000 work-hour analyses, if 200,000 work-hours occur in 250 days what do we get if we divide 14.3 injuries into 250 workdays? The answer is 17.5 days.

Thus in 1989, as an average, one hundred employees worked 17.5 days between Recordable injuries. This means that for a period of near 18 days there were Zero Recordable injuries. This 14.3 rate in 1989 had decreased in 2009 to 4.3, or one injury every 58.1 days; a big improvement. How did this giant reduction happen. It happened one day at a time as employees, leaders and workers alike, across America began reduced at-risk behavior.

So there you have it!

The question is, "can a construction employer beat the

current national average and by how much?"

Since 4.3 is an average, and since some are better the average and some are worse than the average, the answer is a clear "yes" for beating the average. Some do.

The Zero Injury Safety Leadership Concept applied in your company will allow you to achieve safety performance 5 to 10 times better than the national average. For instance in construction some have achieved zero injury with many having results 10 times better than the national average where injury rates are high compared to other industries.

One contractor exceeded 4,500,000 hours in 2003 and 2004 with zero recordable injuries with at least two more contractors working over 2,000,000 hours with zero recordable injury! The 2,000,000 hours is equivalent to 100 employees working 10 years with no recordable injuries. Quite remarkable, I say!

Regarding the above example, the obvious question is, "Does your company measure how many days, or work-hours you work between recordable injuries?"

Many employers do not.

All employee groups work some number of total hours between injuries, whether that injury is a First Aid Case, a Recordable or a Lost Workday Case. So, as a first step, I urge you to begin to measure total hours worked between injuries. Then compare your result of how you are doing to the OSHA/BLS national average for your industry to see how much improved over average your employees can become. Adopt and use the zero injury research information and commit to improve until you too join those who have achieved world-class (<0.20 TRIR) safety performance.

Why do I say <0.20 TRIR is World Class? Here I am looking at a 1,000,000 hour string (one year or multiple years) ending with a recordable injury in hour 1,000,001. You see if an employer works say 600,000 hours ending with a recordable in hour 600,001 the recordable rate will be only 0.33. It requires the 1,000,000 hours to reach the lower TRIR of 0.20. In the past 20 years in the USA I have recorded 15 construction companies achieving this amazing performance and I fully recognize there likely have been more. It is just that with my ear to the safety media pretty much full time I have only heard of 15. In any case it is a remarkable achievement and worthy to be classified as "world class!."

An "often unasked" question regarding Zero Injury is: "Could you give me the logic behind this Zero Injury approach to safety because it seems so unbelievable?"

Read the next chapter for your answer!

CHAPTER 4

THE ZERO INJURY LOGIC

Who wants an Injury?

The first principle question in the safety logic test is this;
"Who wants an injury to occur?"

The answer, of course is "No one!"

Not the worker, not the worker's family, not the employer, not the CEO, not the owner, not the company management, not the worker's supervision, not the worker's co-workers, not the union if the worker is represented, not the employer's client/s, not the community and not government. All are for Zero Injury! That makes up a string of 12 "zeros," all wanting a zero injury outcome. What is the sum? ZERO!

So what is the problem?

The problem, I believe, is found in our past and it seems that some are not convinced they should change from the way it's always been done. We look at our past and see the frustration and helplessness we have felt in trying to reduce employee injury.

With a glimmer of hope we ask; "How can it be possible that some companies work for millions of hours without recordable injuries? If they, in truth, are doing this and with 'no-one' wanting an injury, why do we not change to this zero injury safety leadership process also? And work to make it happen rather than accepting the 'status quo?'"

What are some of the issues to be addressed in changing?"

Setting goals allowing for some number of injuries is one of the "status quo" problems.

Believing Zero Injury cannot be accomplished is one of the "status quo" problems.

Doubting the numbers of those who are successful in eliminating injury for remarkable periods of time is one of the "status quo" problems.

Want to get out of the "status quo?"

The Logic Against Injury Expectations above "Zero"

Second in the logic test is the following question.

"Why not set goals for injury as long as they are stretch goals?"

In my experience some people like to argue that it is unrealistic for the injury goal be to "Zero." "After all," they argue, "we have always had injuries and injuries will continue to happen. So to set a Zero Injury goal flies in the face of reality."

The "goal setting" doubters claim that: "In all reality, injuries will occur, after all, employees are people and they will not follow the safety rules all the time. Therefore we are being very realistic when we set goals for injuries."

Such logic is critically flawed. Allow me to illustrate why.

Let's say you own a small company with 100 employees. Last year your employees experienced five serious lost workday cases. The cost was high, but more importantly you saw the misery and suffering that followed those injuries and as the owner you vowed to do something about it. After consultation with your leaders you decided to set an aggressive goal for the year upcoming of only two lost workday cases.

You then gathered all the employees into a safety meeting. On the agenda was your announcement that you were very concerned, that last year there were five lost workday cases and you wanted them all to know that as the owner you were setting a goal for the following year of only two lost workday cases. All your people were enthusiastic and seemed to buy in on the goal of only two.

Question: On the first workday in January of your new fiscal year, how many of your 100 employees think, as far as you and the goal is concerned, it is OK for them to have an injury that results in a lost workday case?

The answer is, of course, all 100. Each worker thinks that as long as the goal of two is not reached it is OK for a serious injury to occur! And each employee knows, in reality, that one of those injuries can be his or hers should they be so unfortunate!

What have you inadvertently done by setting a goal of "two?" The goal of "two" for the new year has indicated that it is OK for these serious injuries to occur as long as there are no more than two! Not at all the "lets stop serious injury message" the owner was trying to give the employees.

Now, let's take the scenario further. Say that everyone buckles down and indeed your employees work all the way to July without a serious injury: much better than last year. But an injury does occur in July. Now all of a sudden the goal is

37

no longer two but one! Everyone works even harder. Then as "bad luck would have it" another injury occurs in October. Tough times, the goal of two is now all used up.

Question: What is to be the goal for November and December? Reset the goal to one more injury or leave it at Zero as is indicated?

That is correct; Zero has to be the goal for November and December unless you want to change the goal now that it was not reached. Not a likely choice!

Is it OK for the goal to be zero in November through December? Of course it is! If your answer is yes as is mine, then answer this; "Why is it not also OK to have a goal for the entire year of "zero" beginning in January?"

"Zero Injury" Obviated by Setting Injury Goals

Based on the above logic, I believe that a "no injury performance" result is "not at all likely" through the common process of setting goals for injury. For when company leaders set goals that allow for some number of injuries; these goals are set, perhaps inadvertently, acknowledging and thereby accepting that a certain number of injuries will be OK should they occur. This result is the reality even as deep down in our selves we would like no injuries.

> Truism: as employees acknowledge some number of injuries as the goal there also comes the subtle message to all the company employees that injury is permissible and up to the set goal number injury is OK as long as the goal number is not exceeded.

Some may argue that setting goals is not saying it is okay for those injuries to occur. My response is, "then do not set the goals for injury, if you want no injury be brave and ask for that result!

The goal setting subtle message becomes a powerful enemy to the creation of a Zero Injury safety culture because it allows all employees including the CEO the latitude to sacrifice safety in small ways, (mostly sub-consciously) always in the interest of production, schedule or cost as long as the injury goal is not exceeded.

As a result, all employees have reaped a psychological message from the hidden thought that, "As long as the company has not reached our goal for injuries, if I were injured today, my injury would not cause us to go over our goal; therefore my injury would be OK."

Such a hidden and subconscious mind-set undermines the very condition management is seeking; the safety cultural platform where everyone, energetically, focuses their attention on a continuing basis on eliminating at-risk behavior, thus avoiding all injury.

In contrast to setting goals for injury, what is so powerful about the "no injury is acceptable" company position?

Leadership in Safety *"Commitment"*

I believe that Leadership in safety needed to achieve zero injury contains at least ten elements.

First, the leader must have the vision of a zero injury result.

Second, the leader must be able to pass that vision on to those being led.

Third, the leader must have the discernment to staff leadership positions with people who have the capability to lead their group to accomplish the vision.

Fourth, the leader must have visible integrity in leading the Zero Injury commitment.

Fifth, the leader must be willing to ensure the funding is always in place for the expense of the culture change process.

Sixth, the leader must ensure the culture change strategies follow the proven Construction Industry Institute (CII) research findings.

Seventh, the leader must be active in the oversight of the culture change implementation strategies.

Eighth, the leader must be tenacious in achieving the vision.
When the top company leader sincerely makes the "no incident/injury is acceptable" commitment to safety and communicates the expectation for zero at-risk behavior to all employees, including leaders an expected result can come: that result is success!

Ninth, Success comes, however, only if lower management buys-in and reinforces the CEO's commitment, and then only if the workers buys-in and completes the organizational buy-in from top to bottom.

Tenth, along with the above, when the monetary support appears for safety training, safety procedure development, and implementation of the CII Zero Injury safety techniques required to make it happen.

With such a "money where your mouth is" approach the employees begin to believe that, at least for this company, the management commitment is far more than "lip service." Employees see management "walking the talk" daily as they support their safety needs in an example setting manner.

When the employees see this consistency it is then that they, too, become believers and they can, with this demonstrated management support, take all the necessary steps to avoid at-risk behavior.

Further, in a Zero Injury culture, the employees are never badgered by supervision about meeting schedule at the expense of safety: when to do so clearly involves leader at-risk behavior and creates the possibility of an incident/injury.

This can only happen when supervision clearly knows that an incident/injury is an unacceptable tradeoff for an improved schedule.

Safety as a Core Value

With such a pursuit of zero incident/injury it becomes apparent to everyone that when the CEO/COO takes this critical step and clearly announces, "incidents/injury are unacceptable," that "safety has become a core value rather than Priority One."

By definition a "Core Value" does not fall victim to priorities; rather a "Core Value" describes who you are as an individual. "Core Values" may be described as those moral absolutes that are the principal driving forces that all production planning is built upon. Here one could say the desired end Value is Safe Production.

As top management consistently demonstrates the Safe Production message, the company middle management and the hourly personnel, knowing they are going to have to answer to the CEO if an incident/injury occurs, are then free to place total support behind eliminating the all at-risk behavior.

Management employees, who historically may have argued against using a particular safety procedure that was giving them a production problem, now quickly agree on the proper and safer course of action and move on. Why?
Zero incidents/injury is the clear corporate expectation, and no one wants to be found lacking in supporting safety should an injury result. Such companies become our examples for the definition of the word "commitment."

It is in these companies that all employees know that success in the prevention of injury is esteemed above all the other means of success measurement; i.e., cost, schedule, profit margin. It is also in these companies that competitive edge and profit improve!

Zero is the correct approach.

A corporate commitment to Zero Injury sends a distinct and unmistakable message to all employees that any injury is unacceptable. Since "at-risk" behavior is the root cause of incidents/injury the safety training should make it clear that it is "zero at-risk behavior" that will be the key to success.

Note also that such a corporate commitment does not undermine the psychology of those in the workplace as the setting of injury goals does for now all seek to accomplish that which all want first and foremost: Zero Injuries.

Such a commitment raises the expected performance standard for all, supervision and workers alike.

Such a commitment eliminates all discussion in the field about "how safe we need to be to achieve zero injury." All pursue the safest way.

<u>Such a commitment</u> allows all to work together to achieve world-class safety rather than simply some amount better than the National Average.

Think about this; managers and leaders must do something different if different results are expected. The following is a paraphrase of a statement made by Benjamin Franklin over 200 years ago defining "insanity."

"Doing the same thing in the same way year after year and expecting different results."

The workplaces of the world are begging for a new approach to worker safety!

The logic of taking a commitment to zero injury is being proven by dozens of companies worldwide. These are the leaders; the starting place for your new safety culture to come into being is to copy what these leaders do.

Management must lead in this revolution in safety.

A new dream has been born.

Are you going to take part?

One day at a time.

Work forces in the thousands have always been able to work an eight-hour day without injury to a worker; or even a week, or perhaps a month or two. The Zero Injury concept is nothing more than asking a work force to work injury free "one day at a time" for an extended period.

How long, you ask?
Remember, as mentioned above, the 2009 national average for "days worked" in construction with a crew of 100 workers

without a recordable injury was 58 workdays. In 1989 this number was 17 workdays. (days are rounded to nearest full day)

The following is how the number of days is calculated. The Recordable Rate will be used in the illustration.

The BLS/OSHA Recordable Incident rate in construction was 4.7 for 2008. With the 4.8 rate for 100 workers the days between injury calculation is –

$$\text{Days} = \frac{100 \text{ workers x } 2000 \text{ hours per year}}{4.3 \text{ Injuries x } 800 \text{ hours per day}} = 57 \text{ days}$$

This average is 40 days better than it was in 1989 when the average was 17 days.

Companies that have committed themselves to a Zero Injury safety culture are contributing a large part of this improvement.

The Real Truth

During the 20th Century the real truth was that the principal emphasis on worker safety on the job was to "try" to work injury free. I put quotes on the word "try" to explain that 20th century standard was a "small safety emphasis" by today's standard at the beginning of the 21st Century. That "small emphasis" grew to a "large emphasis" by the year 2010.

While achieving Zero Lost Workday Cases was the principle focus during the 1990 to 2000, the effort is now focusing more and more on achieving Zero Recordable injuries. The best records in USA construction now run to over 4,000,000 Recordable free hours. To get a perspective on the significance of 4,000,000 work-hours think of the milestone this way: four million hours is equivalent to a "workdays between recordable injuries" rate for 100 workers of 20 years (5000 work days).

This is nothing less than an amazing accomplishment. Quite a contrast is 57 days to the 1989 OSHA/BLS National average for construction of 17 days, wouldn't you say?

Now there are many more at the beginning of the 21st century that are achieving the 1,000,000 hour mark with Zero Recordable injuries.

These are the world trendsetters, the best of the best.

They lead the Zero Injury believers and achievers!

Believability

Some people like to doubt the zero injury data coming out of the construction industry. However, I stoutly maintain the data are valid. In proof all I can say is that it is very hard, to even impossible, to keep several dozen seriously injured and maimed employees hidden on a large construction project. Look around on these jobs. Talk to the employees; if they give testimony to the truth of the data, it is true.

Unfortunately, one has to recognize that there may be a few who do "play games" with the injury record keeping. These are only kidding themselves. Record keeping integrity is mandatory for a Zero Injury work culture to exist. If the employees see "game playing" with the recording of injury, management loses the worker support so vital to a true Zero Injury culture. But, thankfully, these that manipulate injury record keeping are in a minority.

Construction worker safety in the 21st Century is being pursued in unique and admirable ways. The real truth is, "worker injury has experienced a dramatic reduction through the use of the Construction Industry Institute's Zero Injury research results."

The above are but leading examples. There are many more employers in construction, recognized as a hazardous occupation, that manage their worker safety effort to where worker injury is so rare as to be an uncommon event.

CHAPTER 5

THE ZERO INJURY RESEARCH- SUMMARY

Origin of Zero Injury in the Construction Industry

It was in the late 1980's that the rare safety performance of "Zero Lost Workday Cases" came to the attention of a group of construction industry leaders.

This attention was brought to bear by Ed Donnelly, CEO of Air Products and Chemicals of Allentown, PA. Mr. Donnelly, seeing the results of their own safety management efforts directed at their contractors virtually eliminating serious injury, had a vision that such performance was within the reach of all. Mr. Donnelly then led his CEO peers, all members of the prestigious Business Roundtable, to begin recognizing contractors and owners for safety excellence. It was after Air Products was awarded the prestigious Business Roundtable (BRT) Construction Industry Safety Excellence (CISE) award that Air Product's remarkable safety record in construction became public knowledge.

What was so remarkable? Listen to this. Air Products and Chemicals contracted 2,400,000 total work-hours of construction during 1984, 1985, 1986 and 1987. This was done using over 450 individual contractors across the USA in over 180 projects with a worker injury result of Zero OSHA Lost Workday Cases! And with a Recordable rate of only 2.1, which was some six to seven times better than the OSHA national average in those years.

Comparing Air Products performance to the industry average for those years, we see their approach to contractor safety management avoided some 130 Lost Workday Cases and another 140+ Recordable cases. The question from Air Product's manufacturer owner peers, using the same construction industry resources, was simply put: "How do you do that?"

Another example was found in Winway, Inc., an industrial contractor located in Freeport, Texas, who was also honored with a BRT CISE award during 1988 for having achieved four years with their 600 employees without a lost workday case. This was for a total of 4,800,000 work-hours. Even more remarkable! Again the question arose: "How do you do that?"

Research Conducted

Dr. Richard Tucker, Director of the Construction Industry Institute (CII), Austin, Texas agreed to provided the answer to this critical question. The CII commissioned The Zero Accidents Task Force in late 1989. Dr. Roger Liska, of Clemson University, the Academician serving on the task force, performed the actual research in cooperation with the Task Force. The research project required nearly four years of effort by the task force members. The author served as Task Force Chairman.

This Task Force Mission was aimed at determining how some employers were able to work these millions of hours without serious injury to their employees while others found it difficult to have less than the then six ("the national average") lost workday cases per 200,000 hours worked. They studied 25 projects, eight of which were experienced zero lost workday cases.

The August, 1993 Task Force report concluded that there were five critical safety techniques being used by these contractors who were successful in eliminating Lost Workday Cases. These were:
- Pre-project/Pre-task safety planning
- Safety orientation and Training
- Written safety incentive program
- Alcohol and substance abuse programs
- Accident/Incident investigations

More details of the 1993 research report are given in Chapter 9.

Industry Performance Results from the 1993 Research

In the years following many owners and contractors applied these techniques with gratifying success. Serious injury rates plummeted, schedules were improved and costs reduced.

During the years between 1993 and 2000 many contractors began experiencing "Zero Recordable" results on some projects. Thus it seemed that if the Zero Injury techniques were applied with certain other key safety management ingredients, it was possible to complete projects with Zero Recordable injuries.

When the initial research was completed in 1993 and published, and the members of CII began using the Zero Injury techniques the performance results of the CII member companies, when measured by OSHA/BLS Recordable rate, dropped from 7.20 in 1989 to 1.02 in 2001.

By the year 2000 many projects were being completed with Zero Recordable injuries. It was clear that newer methods of managing a work force for the elimination of injury had evolved.

Just as in 1989-1993, some were managing project safety with Zero Lost Workday Cases; by 2000 many were managing project safety with Zero Recordables. This result was obviously very gratifying for those following industry safety performance trends.

With these successes and the time lapse since the 1993 research report it seemed mandatory that follow-up research be accomplished. Thus in 1999 CII once again commissioned a research task force.

Follow-up Research Conducted

Ten years after CII commissioned "The Zero Accidents Task Force" a new task force was commissioned titled "Making Zero Accidents a Reality." This task force reported its research results in August 2001 and August 2002.

This task force enlarged the number of critical safety management techniques to nine. These were:
- Demonstrated management commitment
- Staffing for safety
- **Safety planning**
- **Safety training and education**
- Worker participation and involvement
- **Recognition and rewards**
- Subcontractor management
- **Accident/incident reporting and investigations**
- **Drug and alcohol testing**

The 1993 Research results are shown in bold type. More details of the 2001/2002 research reports are given in Chapter 10.

The Questions

First: Will the principles found in the CII research apply to a workplace that is not in construction?

The answer is a clear "yes" with appropriate minor modifications.

The second question on the mind of those who do not yet walk the "Zero Injury high ground" is simply this:
"How do they do that?"

Translated –
"If I can find out how they do that then perhaps we can also!"

If you are one of these people asking this question and are sincere in this interest, and wish to spend a few hours boning up on this topic, then you will find that there now is "a clear pathway to zero injury."

The process described below that can change your working culture is applicable to any work place, be it transportation, manufacturing, mining, or simply an office staff. The process is universally applicable.

The third question on the mind of the interested is: "How do we get started?"

Well, if you have read this far, perhaps you already have started. It's up to you.

The fourth question is: "What is involved, what do we have to do?"

This book is your introduction and guidebook to implementing the Zero Injury concept in your company. This process could be named "Installing the Zero Injury Safety Leadership Concept." Read this guidebook and apply what you read. It is not a quick fix.

It is journey up the "Doing the Right Thing Highway" to higher ground and as you go determine who the successful companies are and learn from them. The larger the employee population hence the safety culture the longer it will take. For employers with as many as 1000 employees and more it can take five to seven years.

CHAPTER 6

CORPORATE LEADERSHIP IS REQUIRED

The CEO is Key

Corporate Commitment

The key research findings was: <u>In the zero lost workday injury companies the CEO always had a key operating safety expectation placed before the company management.</u> It was, paraphrased as follows -

> *"We will do our work without an injury. It is my (the CEO's) belief that all injury can be prevented and it is my expectation that there be no worker injury on our projects. And if an injury does occur it will not be viewed by me as acceptable performance! And I personally will be involved in determining how management failed. We will not set goals for injury! Our commitment is to ZERO Injury! This is not a statistics management effort. Rather our commitment shall be a complete devotion to the elimination of unsafe behavior by all employees, management and workers alike."*

A CEO Speaks

I personally heard the CEO of a large construction company located on the West Coast tell of his experience regarding his attention to safety. He related that when he remained focused on safety he saw injury rates come down but if his attention was diverted for a time the injury rates rose.

Conclusion; CEO attention is mandatory!

The CII research found that in companies such as the examples listed above that, indeed over time, a successful "unsafe behavior will be avoided" culture had been established and was pervasive throughout the organization, reaching effectively into the hourly ranks.

Some realists will want to ask, "What is the difference between the above position of adopting a commitment to Zero Injury and a more "normal" approach of, over time, setting increasingly more stringent goals for injury frequency? Doesn't the latter approach work just as well?

They add, "It is not realistic to expect zero injury is it; after all, injuries are always going to happen aren't they?"

The author is the first to admit that, given we are working with fallible humans at-risk behavior will at some point be engaged despite all out efforts and an injury will occur. But remember the statement given above on page 21, "The fact that injuries occur does not mean they must occur!"

Many employers have proven this statement to be true. These have simply learned to use CII proven approaches to managing safety, involving the employees in unique ways to the point where employees "buy-in" the frequency of "at-risk" behavior subsides and injury becomes a very rare event.

A corporate commitment to the concept of "zero injuries" carries the heavy responsibility of not only "talking the safety talk" but also "walking the safety talk."

"Walking the safety talk" is another way to say "Demonstrated Management Commitment." It means doing whatever it takes to work as a team to prevent the next "at-risk" behavior at the worker level and also to prevent the not so often focused on

54

next "at-risk" behavior by a leader! Yes many times the root cause of an injury can be traced back to an action or lack of action by a leader.

For those who have had such a commitment and failed from my personal experience they failed because they did not "know how" to create a working culture where worker injury is rare to non-existent.

Before the CII zero injury research was accomplished few knew how but many were trying.

In a zero injury culture management actions "must" unequivocally parallel management's spoken word. The research found a direct correlation that safety training, the fourth item on the above list of nine, was a significant component in creating the possibility for Zero Injury to become reality. Thus the management actions taken must include emphasis on the safety training of the worker.

"Training" has proven to be a vital complementary element that yields progress toward zero injury.

Over the past 15 years and many, many zero injury workshops with hourly crafts personnel as well as leaders "employee safety training on the zero injury concept" has proven to be among the top three "to-do" priorities cited by the workshop attendees.

Once all required management actions are in harmony with the "talk", with zero injury training done, injuries caused by the "uninformed worker" disappear and injuries caused by the "careless act" begin to disappear.

Once the "zero injury" concept is believed and accepted by the employees, injuries caused by worker loyalty disappear. Yes, if a loyal worker does not have a clear message from management that "at risk" behavior is not acceptable then some will take chances in the interest of being good and loyal and productive workers. In a Zero Injury culture taking chances with safety is not condoned.

It is true, many injuries do occur out of "chance taking." Chance taking arises when the worker thinks s/he is going that extra mile to shorten the time (taking a shortcut) to accomplish a task, or when the worker attempts to prevent or reduce the damage once an accident is in the process of occurring.

One of the significant safety gains that comes with the commitment to "zero injuries" is that the employees know that it is no longer O.K. to take chances out of loyalty to the employer that might result in injury.

Zero Injury as a Concept has Reached Critical Mass

Achieving excellence in safety performance through the Zero Injury concept is becoming more and more the accepted norm in owner and contractor companies in the USA. I predict that the Zero Injury Concept movement has reached critical mass and will continue until it becomes the sought after safety culture in most of American Industry.

The zero injury concept has already reached the status of "the cultural norm" in several regions of the USA. Among these are the Texas Gulf Coast Region, because of the Petro-chemical industry that have led in setting the expectation that zero injury was what was desired. The San Francisco and Los Angeles region due to the influence of companies such as Chevron, and Shell Oil and the East Coast region for those contractors that work for E.I. du Pont who have long been the industrial leader in eliminating injury from their workforce. Steel companies such as US Steel made great strides in the early 2000's leading contractors in the Zero Injury Concept. And there are signs that the electric power industry is beginning to capitalize on the savings coming from a zero injury workforce. A major effort was launched in Canada to publicize and encourage contractors to embrace the Zero Accidents notion.

In this day, 2011 an internet search for the words "zero injury" turns up over 10 million sites.

Setting Expectations at Zero

There is no substitute for the concept of setting the ZERO safety performance expectation for your employees! It should be the ongoing plan of each employer that employees working for them, directly or through a subcontractor, will "know" that the expectation is for "zero" injuries and to accomplish this we will not engage in at-risk behavior.

How you begin the quest for zero injury is critical. In the beginning the Zero Injury expectation needs to be believable by your employees. To ensure believability many are first setting their injury commitment to Zero Lost Time Cases. Wait, you say, all the time you have been talking about zero injuries and now you are saying set the commitment for zero lost time injuries.

There are two reasons for this approach. The first, if your employees have never been exposed to the zero injury concept, they will typically "laugh in your face" if told the commitment is to zero injury. The credibility of the Zero Injury Concept must be established with your employees over time. The second reason is the concern about the potential for hidden injuries. Especially in a work force that for the first time is given the expectation to work injury free. Some begin by recognizing zero lost time injuries; then after achieving this performance for extended periods, change to zero recordable injuries as the recognition norm.

Around the USA in the year 2001 there were many examples of construction work sites where millions of hours were worked with no resulting "lost time case" injuries. In 2011 the number of projects that are completing at Zero Recordable free are becoming abundant. In achieving this level of performance, human suffering is largely eliminated and the cost savings are very significant.

More and more participants in the construction business, both owners and contractors, are coming to realize that "zero" lost time injuries is achievable. And they now are taking that next step; telling their employees that "zero at-risk behavior" is the new expectation.

As a CEO, letting the employees know of your "zero" injury expectation requires another sobering commitment; it is safety support at the job site through a well-planned and executed safety program that embraces the CII research.

The Competitive Edge

Even though an employer's overhead for Workers' Compensation Insurance is passed on to the user of product or service, those who maintain a near zero injury level of excellence in safety performance have a competitive edge on those less skilled in eliminating injury in the workplace. The lower costs and improved productivity arising from the improved safety performance of a zero injury result flow to more competitive bidding and increased profits.

CHAPTER 7

INJURY RATES AND THE COST OF INJURY

The OSHA/BLS Injury Rates

It is a fact that over the past 20 years injury rates in American construction have decreased significantly. Would I say that this decrease is totally explainable by the advent of the Zero Injury Concept?

No, I would not make such a statement, for I believe this reduction has been a result of the increased cost of injury and many other even more important positive actions within the industry and government such as I list below in no particular order:

1. OSHA attention to the safety standards.
2. Broader attention by Labor Unions to safety issues.
3. The simple cost of injury as driven by Workers' Compensation insurance.
4. Recognition of the significant indirect cost of injury.
5. Attention by the Business Roundtable Construction Committee in recognizing safety excellence.
6. Broader attention by Contractor Associations to recognizing safety achievements by their members.
7. The results of the Construction Industry Institute research into the Indirect Costs of Injury.
8. The results of the Construction Industry Institute research into projects with Zero Accidents.
9. The advent of a number of Behavior Based Safety

consultants who deal with the leading indicators or precursors to injury.
10. The attention paid by the Safety Associations and Councils such as the American Society of Safety Engineers and the National Safety Council.

As a result of the above partial list of contributors, injury to the American construction worker has steadily declined since 1988.

Importantly, as a consequence, contractors and owners have saved money in medical costs and on the even more debilitating indirect cost of worker injury.

The solid evidence is in!

Injury rates as measured by OSHA/BLS tell the story.

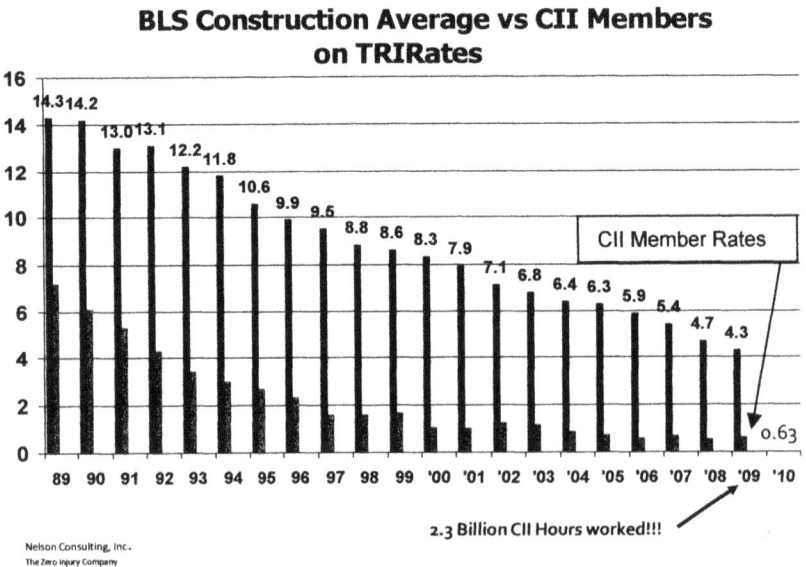

BLS Construction Average vs CII Members on TRIRates

Nelson Consulting, Inc.
The Zero Injury Company

The above data do not extend beyond 2009 due to the lag in OSHA/BLS obtaining and processing the mass of data.

While this decrease is in itself gratifying, there is also an amazing reduction in injury rates by the member companies of the Construction Industry Institute (CII). These are the principal companies that have applied the Construction Industry Institute Zero Accident Task Force research results to their projects.

Note the 1993 research results have now been published eighteen years and most of the CII member companies have made significant improvements to the safety techniques revealed in the task force data. The following two charts comparing CII data to the OSHA/BLS data tell this amazing story.

20 year trend BLS National Average vs CII Member DART Incidence Rates

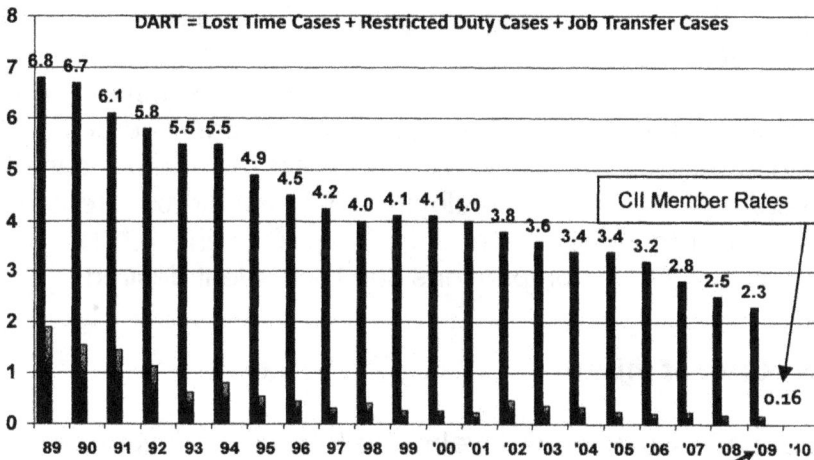

DART = Lost Time Cases + Restricted Duty Cases + Job Transfer Cases

CII Member Rates

Nelson Consulting, Inc.
The Zero Injury Company

2.3 Billion CII Hours worked!!!

3

The 2009 CII data represent over 2.3 Billion hours worked that year; thus, the impact on the national average is significant.

Before we leave the above chart I would like to point out one more amazing fact that is displayed. The CII member DART rate for 2009 is less than the 0.20, (the world class threshold for recordables.) This 0.16 DART IR means that on the average the CII members are experiencing a DART on a frequency of over 1,000,000 hours; in fact the number is one DART every 1,250,000 hours. AMAZING when in 1989 the more severe Lost Time incident rate was one each 29,400 hours! "Wow," is all I can say.

OSHA Makes an Adjustment in 2002

Beginning in 2002 OSHA implemented new rules on how to make a decision on whether an injury is recordable or not. In short, the determination rules were tightened. This change will caused an increase in the averages for each employer and for industry averages in general. Whereas in the past each year OSHA required an annual report by each employer in the "OSHA 200 Log" format, the new form or format is named the "OSHA 300 Form or Log." Therefore, all historic records will not be exactly comparable to the current records. Of course, an employer could take the past records, alter them to bring all "record of injury" to the "OSHA 300 standard." Thus, comparisons can be made if desired.

The Cost of Injury

There are two ways to develop the total cost of injury. One is by using the actual loss data; the other is to analyze the underlying costs of carrying Worker's Compensation Insurance.

The direct costs of injuries as driven by actual injury costs vary widely between contractors because of the difference in how cases are managed. Contractors often make no attempt to get the injured back to work in a productive and medically approved work activity. In such cases the costs can get very high indeed.

However, the average costs of an injury, where the employer works actively with the medical provider to get a restricted work release, are much reduced. An employee back at work and productively engaged in activity that does not jeopardize his/her physical condition is known to heal more rapidly and thus is back on full duty sooner.

One thing I have noticed over the years is that as those contractors successfully utilizing the CII Zero Injury safety techniques watch injury rates fall each year, they also observe that the severity of the injuries that do occur is also much less. With reduced severity there is the obvious reduced cost.

Determining the average cost of injury as driven by Workers' Compensation Insurance premium can give one a notional look at injury cost that can be used to motivate top management to pay increased attention to safety.

Calculating the Average Cost of an Injury

In a broad sense this method takes the total cost of Workers' Compensation Insurance and divides by the number of injuries for the same period.

To get this cost we need to calculate the average premium for a construction situation. The premium is generally calculated by the following formula: (simplified)

Premium = Payroll x Manual Rate x EMR

Your immediate question may be, "What is EMR."

The short answer is, it is the ratio between your actual injury losses to your expected losses. The higher your losses due to injury the higher the EMR and a high EMR is a costly thing to have to endure. I have heard of EMR in the range of 3.00 and I have heard of EMR in the range of 0.20. Obviously the one you want is the lower one.

Your next question may be, "What is Manual Rate?"

The Manual Rate is the Workers' Compensation Insurance rate per $100 of payroll. These rates are different by trade and by state. If two rates are the same in two different states it will be a coincidence. These rates in construction range from $2 to $4 dollars to as high as near $100 in the more injury prone trades in one or two states.

Using 100 workers let's assume the EMR is 1.00 and the payroll is $8,000,000. This payroll would represent 100 workers at a $40.00 wage rate for one year. This wage is applied to 2000 work-hours per employee per year.

The serious injury (DART case) rate per 100 workers was 2.30 in the year 2009. My experience has shown that about 80% of all injury costs are attributable to DART Cases.

Let's say the national average "16 trade" Workers' Compensation "Manual Rate" for construction in the USA is about $5.00 per $100 payroll and EMR = 1.00.

Thus: Nominal Premium = (8,000,000/100 x 5.00 x 1.00 = $400,000 for 100 workers.

The OSHA/BLS total recordable rate for the year 2009 was 4.30. By subtracting the 2.3 from 4.3 we get 2.0 other Recordables (4.3 − 2.3 = 2.0). If we estimate the cost of each non-DART at $2000 and multiply by 2.0 we get $4,000. Then by subtracting $4,000 from the $400,000 we get $396,000 as the cost of the 2.3 DART cases. If we then divide $396,000 by "two point three" (2.3) we get $172,173 premium per each lost workday case.

Remember this $172,173 would be the national average premium cost paid per DART case. Also recognize the above calculation does not consider other negotiable cost factors that may be included in a Workers' Comp. policy, such as insurance carrier discounts, etc., which reduce overall premium for the affected contractors. Such factors can lessen the cost for these if the employer actively seeks the best rates from a number of carriers. .

So right away one can see the desirability of an EMR around 0.20 (the best I have seen). With a 0.20 EMR the total premium for such a contractor falls from $396,000 to $79,000 per 100 workers. Conversely the premium rises for an EMR of 3.00 to $1,188,000, a $1,109,000 difference. By the way, this saving does not include the reduced indirect cost which will be larger than the premium cost just calculated.

Calculating the Total Cost of an Injury

To get the total cost, the average cost must be adjusted to include the average indirect cost of an injury. CII research found that the indirect costs of the average injury run 2 to 20 times the direct costs (medical and indemnity). But for the sake of a conservative approach, let's assume only a "one to one ratio."

To find a pure indirect cost out of the Workers' Compensation premium we need to reduce the per injury premium above by the amount of insurance carrier overhead and profit then add the remaining amount. See example calculation following.

Let's say that carrier's charge an average of about 30% for

overhead and profit. This reduces the $79,000 by 30% or $23,700 leaving the estimated pure indirect costs at $55,300. To honor our 1 to 1 ratio we must multiply the $55,300 by 2.0. This brings the total cost of a DART to case to $110,600.

Explaining the Cost Result

If this $110,600 looks high, it is. It is high for many in 2011 because in a number of states the manual rates are higher than they need be so the carriers will give deep discounts on premiums. Notice, I said in a number of states. Counter balancing this, some states have such a lenient Workers' Compensation administrative process that it results in significant misuse and fraud, which has become all too common. In those states this dramatically drives up the costs to the employer.

The costs look high because for many contractors the total of $110,600 is what I term the cost of negligence. This could be the cost where a contractor paid no attention to their Workers' Compensation coverage or to getting the injured back to work. In such cases, then, these higher numbers may well represent the true cost. In any case such a number is sobering to see!

With a successful Zero Injury initiative in place and functioning well, coupled with an alert case management process utilizing the latitudes allowed, to ensure return to work when appropriate, these costs are much lower. It is the experience of the CII member companies that the average cost of a DART case will run around $40,000 not including indirect costs. Of course this lower cost also varies from contractor to contractor and will also vary from region to region as medical costs, wages, and state partial wage vary.

Again, please note that, the estimated $110,600 is the national average cost assuming no carrier discounts, and as mentioned above, of doing no case management; in which case, I term the above figure the cost of negligence. Sadly there are some who fall in this category.

Chapter 8

Zero Injury Return on Investment

The Basics

Would Zero Injury be of more interest if I told you that it is routine for successful users to get a 300-500% ROI on the cost of implementation and evergreen maintenance of the essential zero injury research based safety culture building techniques?

There are a number of questions regarding data gathering surrounding the effort to calculate the ROI obtainable when changing a safety culture from "injuries happen" to one that embraces the Zero Injury Concept (and ultimately achieves Zero injury on a scale with world class performance.)

In today's age we all know the costs of injury are very high. But little do we realize that it is a lot higher than we think. The "direct cost of injury" (that covered by Workers' Compensation) is given to us in our Workers' Compensation management processes when we audit our "Loss Run." By adding to our "injury losses" the carrier's charges and our hidden "losses in efficiency," and then calculating an ROI on the expense required to reduce injury frequency, one can see that the undeniable financial result is "pure gold" to the bottom line. Add to this the undeniable fact that you have done the right thing then who would argue that to strive for zero injury does not make good business sense?

The "losses in efficiency" flowing from employee injury are calculated by using the CII research based ratios between Injury Losses and "losses in efficiency, termed indirect costs." This research performed by CII in 1988, gave the industry hard numbers on the indirect cost of injury. Yet employers are seldom, if ever, using these data to generate enthusiasm for Zero Injury as a workplace norm through an ROI calculation.

What if a company took their actual "direct cost" data and combined it with the CII research based Ratio between Direct and Indirect cost, and then combined this information in a calculation to determine ROI for the added money spent while in the process of embracing the Zero Injury Process?

My experience reveals that the "pay-out" from reduced injury on successfully applying the Zero Injury principles is not a paltry 10-25% but an amazing 300-500%. I ask the reader; "Is there a quicker way to increase profit than this?"

Company leaders work endless hours to develop business, and then more hours preparing and performing the work, all more often than not, for a slim and optimistic 5 to 10% profit margin. They do this, while right in the midst of the management process, an opportunity exists to get returns on investment of 500%! Since we are talking employee safety, of course, this opportunity rests in how safety in managed.

If you are one of those that are managing safety the same way year after year, with near average unsatisfactory results, then you are the one that will gain the highest ROI on spending the money to install a Zero Injury Safety Culture in your company.

This author has applied specific data (real data) and the ROI result is astounding! Would you believe a 300% return is normal? This is not just in the first year of the calculation.

Once near Zero Injury is achieved the 300% will be there as long as the application of the Zero Injury concept yields world class injury statistics, year after year after year.

Learning this should be an "eye-catching" revelation to company leaders and a wake up call to drive safety success to world class. The Zero Injury effort is most rewarding for those who enjoy performance that are near or over the OSHA/BLS national averages. The closer you are to Zero Injury when you start, the less the ROI on Zero Injury calculation effort will yield. But even if you are down in the 100% ROI range, it cannot be considered as anything other than a "brilliant investment of time and energy."

What is World Class Performance

We touched on this in Chapter 2. World-class performance is working with no injury: Zero OSHA Recordables. There are many companies that achieve zero injury every year. Some have done so with millions of hours worked. It is now common for large and some small companies to accumulate runs of 1,000,000 work-hours with no (Zero) OSHA Recordables injuries. This is world class. You will be world class when you do the same. And you can.

For industries where lost time injury OSHA frequency rates are high, above 2.00 up to 6.00, the pay out is largest. If you have one lost time injury each year and it costs $100,000 and you spend $50,000 to eliminate that injury you are getting a return of 200%. Getting these numbers, in exact terms, is the reason most do not attempt to do an ROI calculation. But in all reality these numbers can be obtained quite easily.

CHAPTER 9

THE 1993 CII
ZERO ACCIDENT/INJURY RESEARCH
IN DETAIL

The 1993 Research

The Construction Industry Institute (CII) affiliated with The University of Texas at Austin, took on a research effort in 1989 looking into the occurrence of "Zero" Lost Workday Cases on construction projects. In 1987, it became public knowledge that some contractors in construction work, the most dangerous of all industries in the year 1987, were working millions of hours without OSHA Lost Workday Cases. Industry leaders asked the obvious question.

"How do they do that?"

Leaders in The Construction Industry Institute volunteered to undertake a research project to determine the answer. The Zero Accident Task Force was formed in 1989 and finished in 1993. Emmitt J. Nelson retired Shell Oil Company Construction Relations Manager served as Chairman. A total of 16 members including Dr. Roger Liska of Clemson University made up the group. The purpose of The Zero Accidents Task Force was:

1) To show owners and contractors how to achieve zero accidents on construction projects.
2) To convince management of the value of an effective

71

safety program, through research, by identifying techniques most successful in achieving zero accidents.

The research was conducted through interviews at 25 construction projects. Seventeen of these projects had an average Lost Workday Case Incident Rate of 0.25. Another eight projects had an average Lost Workday Case Incidence Rate of 2.5, ten times worse.

In order to obtain viable research information three data sets were obtained.

The first data set was the result of interviewing over 400 hundred Trades people by asking each to name the three most important safety techniques being used.

The second data set was the result of asking Trades people and leaders alike for the amount of time being spent on 17 critical safety techniques. The following two questions are examples.

"How much time do you spend in safety meetings?"

"How much time do you spend in Pre-Task Safety Planning?"

The third data set resulted from determining which of a list of 170 commonly known safety techniques were being used by each project.

Out of these data the final research findings were assembled.

The 1993 Findings

After nearly four years of work the task force reported, in late 1993, the top five principal safety techniques with important sub-techniques being used by those contractors achieving Zero Lost Workday Case Injury:

The Employer Safety Guide Book
to Zero Employee Injury
Third Edition

1. Safety Pre-Project/Pre-Task Planning
   a. Pre-Project planning
      i. Safety goals (not injury goals)
      ii. Safety person/personnel
      iii. Pre-placement employee physical evaluation

   b. Pre-Task planning
      i. Task Hazard analysis
      ii. Task training

2. Safety Orientation and Training
   a. Site orientation
   b. Owner involved in orientation
   c. Safety policies and procedures covered
   d. Project specific orientation
   e. Formal safety training

3. Written safety incentive program
   a. Cents per hour for the workers
   b. Spot cash incentives used with workers
   c. Milestone cash incentives given to workers
   d. End of project incentives given to workers

4. Alcohol and Substance Abuse Program (ASAP)
   a. Pre-employment Screening done for alcohol and drugs
   b. Screening conducted at random
   c. Inspections for contraband conducted
   d. Post accident screening done for all employees
   e. All project contractors have ASAP's

5. Accidents/Incidents Investigations
   a. Near hit Incidents investigated
   b. Near hits are reported to home office
   c. Accidents without injury investigated
   d. Project accident review team established for all accidents or incidents
   e. Project work exposure hours and safety statistics reported to home office

Techniques - Time Quantity versus Quality

The research data show that quantity of time spent on selected techniques is not as important as the quality of time spent. An even more significant point is this; in an evolving safety program the selection of the specific safety technique to spend time on becomes more crucial than how much overall time is spent on it.

As an example, time spent in pre-project/pre-task planning results in earlier hazard identification and more hazard control and elimination, therefore, the project will require less hazard discussion and protection. This would result in less time spent on other techniques, such as safety meetings and accident investigation.

Of paramount importance was to, somehow, obtain the buy-in of the Trades people. It was found that, unless they are convinced that management was indeed serious about achieving a zero injury result, the actual result was less than satisfactory.

1993 Research Task Force Conclusions

The task force research provided the following conclusions regarding achieving zero injuries.

*Zero Lost Workday Cases was being achieved on all types of construction and maintenance projects.

*Zero Lost Workday Cases was being achieved by contractors, small and large, and on many projects, even large projects with several million work-hours.
* Zero Lost Workday Cases was being achieved in all project labor situations: i.e., union, merit or non-union.

*Joint owner and contractor senior management devotion to

"ZERO" is key in achieving this level of performance. Setting this expectation for project workers at all levels is a vital first step.

*An effective safety program producing "good" safety performance must contain a broad base of essential safety techniques similar to those contained in the 170-technique list.

*Attaining zero injuries is significantly more likely on Projects that apply the Five High Impact Zero Injury Techniques that were identified.

*Success in eliminating accidents is not guaranteed by use of the Five High Impact Zero Injury Techniques alone.

*Quality of effort is strongly suggested by the "Time Spent" data as a vital ingredient in reaching zero injury.

1993 Research Task Force Recommendations

The task force offers the following recommendations:

*Adopt the "Zero Injury" philosophy, beginning with the chief executive officer (CEO) who sets the expectation that worker injury is unacceptable on all work.

*Create a culture where all employees, at all levels, accept ownership of the safety performance objective of "Zero Injuries." The CEO sets the expectation and empowers all employees to do what is necessary to reach zero injury performance.
*Recognize that profit lost through worker injury is not covered by insurance.

*Institutionalize a comprehensive basic safety process using the 170 safety techniques identified by the research.

*Establish specific contract requirements defining the roles and responsibilities of all parties (companies) involved in reaching zero injury.

*Define, explicitly, the safety responsibilities and authorities for all project personnel (leaders and workers) involved in the project.

*Implement The Five High Impact Zero Injury Safety Techniques.

*Recognize quality of effort is more important than time spent, as techniques are implemented.

*Understand that the high cost of Workers' Compensation is driven by worker injury and that achieving zero injury performance is a key component of responsible management for profit.

*Conduct a safety assessment.

*The owner should be an active participant as the project embraces the "Zero Injury" philosophy.

*Ensure that subcontractors are active participants.

The Application of 1993 Research Data Leads to Zero Recordable Performance

Even as the 1993 research was being conducted it was apparent that safety performance in the construction industry was experiencing a step change of improvement. After the research results were reported, numerous CII member companies began implementation of the Zero Injury principles.

Soon these owners and contractors were requiring use of the Zero Injury techniques by all those working on their projects. As this cadre of contractors and owners enlarged, the movement toward adopting the Zero Injury Concept gathered momentum. It was at this stage that the technique of involving all employees in the Zero Injury process found broad support. It became immediately apparent that employee involvement took the frequency of injuries to a new lower level; Zero Recordables. This was apparent, because many contractors began experiencing Zero Recordable injuries for longer and longer periods of time.

By 1997 many in the construction industry saw that follow-up research was going to be necessary in order to firmly establish the extended family of Zero Injury Techniques that were producing these Zero Recordable results. Thus in 1999, CII launched the next research effort, "Making Zero Accidents a Reality." The 1999 beginning resulted in new results in 2001.

Chapter 10

The 2001/2002 CII Zero Accident/Injury Research In Detail

From Zero Lost Workday to Zero Recordable Cases

Construction industry safety performance progress was solidly verified when the researchers noted that among the 38 projects there were four that were working at the Zero Recordable level when the research was conducted.

Though in 1993 the recordable incident rate on only one project was below 1.00, the Zero Recordable performance was not being achieved with sufficient regularity and length of time to notice. The best in the 1993 research project group of 25 had a Recordable Incident rate of 0.87, which was excellent for the time.

The 2001 Research

Almost exactly a decade after the original Construction Industry Institute (CII) research began in 1989 (results released in 1993), CII once again, in 1999, commissioned a task force to examine recent successes in working in injury free in construction. These results were released in 2001. This time, however, the task force would examine the frequent occurrence of "Zero Recordable Injuries" on construction projects. The "Making Zero Accidents a Reality Task Force" was formed in 1999 with John Mathis of Bechtel as Chairman. A total of 14 members including Dr. Jimmie W.

Hinze of the University of Florida made up the group. The purpose, as issued by CII leaders, was:

- Make zero accidents a reality through research and identification of current Zero Accident best practices that have proven results across a broad spectrum of the construction industry.

-

This mission was later expanded as follows:

- Develop a communication and education component to assist in understanding and implementation of best practices that support a Zero Accidents culture.

The task force conducted two separate studies:

- Large construction firms and
- Large construction projects

Questionnaires were sent to the Engineering News Record 400 largest contractors. There were 102 responses. Detailed interviews were conducted on 38 North American construction projects ranging in size from $50 million to $600 million. The 38 projects covered the following six types of construction:

Petrochemical – Industrial - Public Works
Transportation - Hotel - Commercial buildings

The 2001 Findings

As given above the task force found there were nine "Best Practices" that were critical underpinnings of a Zero Injury culture.

These were:

- **Demonstrated management commitment**
- **Staffing for safety**
- Safety planning
- Safety training and education
- **Worker participation and involvement**

- Recognition and rewards
- **Subcontractor management**
- Accident/incident reporting and investigations
- Drug and alcohol testing

Those not in bold type were the five 1993 research results. Clearly the 2001 research clarified the need for appropriate numbers of safety personnel, and the absolute necessity of management to demonstrate their commitment. Added by the 2001 results were the extremely important message of employee involvement. It is clear that ways to obtain Employee "buy-in" and support is absolutely essential. Pay attention to sub-contractor management, insuring that all contractors on the project are combining their collective skills in the avoidance of unsafe conditions and unsafe behavior.

It is when all workers see this total and coordinated commitment that they actually begin to believe that their employers are really serious about this subject called "safety." With this belief, the advent of injury free work finds its fruition.

Technique Relative Impact

A number of specific sub-techniques were examined seeking to determine the impact each technique had on the reduction of Recordable Injuries. In each case the Task force compared those projects using the technique against those who were not using the technique and reported the average Recordable Rate for each class.

Demonstrated Management Commitment:
Top management participates in investigation of Recordable injuries.
- Participates in every injury – RIR = 1.20
- Participates in 50% or less – RIR = 6.89

Company President/senior management reviews safety performance record.
- Yes – RIR – 0.97
- No – RIR = 6.89

Frequency of home office safety inspections on the project.
- Weekly/Bi-weekly – RIR = 1.33
- Monthly/Annually – RIR = 2.63

Safety Staffing
Number of workers per safety professional
- 50 or less – RIR = 1.33
- Over 50 – RIR = 2.35

To whom does safety representative report?
- Corporate Staff – RIR = 1.38
- Project Line – RIR = 2.41

Safety Planning
Does the project have a site-specific safety program?
- Yes – RIR = 1.76
- No – RIR = 5.43

Are Pre-task meetings held?
- Yes – RIR = 1.04
- No – RIR = 2.67

Safety Training and Education
Is safety training a line item within the budget?
- Yes – RIR = 1.38
- No – RIR = 2.63

Does every worker on-site receive a safety orientation?
- Yes – RIR = 1.76
- No – RIR = 5.72

Is format of safety orientation formal as opposed to informal?
- Yes – RIR = 1.51
- No – RIR = 3.80

Do workers receive at least 4 hours per month of safety training after orientation?
- Yes – RIR = 0.94
- No – RIR = 2.79

Do superintendents and project managers receive at least 4 hours safety training per month?
- Yes – RIR = 1.07
- No - RIR = 2.00

When are tailgate meetings held?
- On Mondays – RIR = 3.25
- On Tuesdays/Wednesdays/Thursdays – RIR = 2.00

The Employer Safety Guide Book
to Zero Employee Injury
Third Edition

- Daily – RIR = 1.0 (When Pre-task safety planning is used.)

Worker Involvement and Participation
Are safety perception surveys conducted on the project?
- Yes – RIR = 1.33
- No – RIR = 2.82

Do management and supervisory personnel receive behavior overview training?
- Yes – RIR = 1.38
- No – RIR = 2.82

Does a formal worker-to-worker behavior observation program exist on the project?
- Yes – RIR = 1.38
- No – RIR = 2.82

Does the total number of safety observation reports filed on the project exceed 100?
- Yes – RIR = 1.01
- No – RIR = 1.93

Recognition and Rewards
Does the project have a formal worker incentive program?
- Yes – RIR = 3.20*
- No – RIR = 2.05
  * This incentive was one of a single large prize at the end of the project.

Is recognition incentive based on zero injury objectives?
- Yes – RIR = 1.33
- No – RIR = 3.29

How often are recognition incentives given to workers?
- Weekly/bi-weekly – RIR = 1.33
- Quarterly – RIR = 3.29

Do family members attend safety dinners?
- Yes – RIR = 0.18
- No – RIR = 2.35

Are field supervisors evaluated on safety?
- Yes – RIR =2.00
- No – RIR = 8.89

Subcontractor management
Are subcontractors required to submit site-specific safety plans?
- Yes – RIR = 1.37

- No – RIR = 3.83

Do all subcontractor workers attend a formal standard safety orientation?
- Yes – RIR = 3.30
- No – RIR = 5.33

How frequently do subcontractors hold safety meetings/pre-task?
- Daily – RIR = 1.04
- Weekly – RIR = 2.45

Are there sanctions for subcontractor non-compliance with safety standards?
- Yes – RIR = 1.43
- No – RIR = 5.35

Accident/Incident Reporting and Investigation
Number of near hits recorded on the project?
- Over 50 – RIR = 0.57
- Under 50 – RIR = 2.35

To what extent are recordable incidents investigated by top management?
- Every injury – RIR = 2.00
- 50% or less – RIR = 5.60

Results of Implementing Best Practices
- Jobs that implement most – RIR = 0.17
- Jobs that implement a few – RIR = 3.84

Worker Involvement and Participation Interventions

During the eight years between 1993 and 2001, a safety technique known as Behavior Based Safety (BBS) that involves the craft personnel had dramatically risen in popularity. At this juncture, as far as I know, only one of the safety consultant advocates of the BBS process have pushed for a Zero Injury mind set. Instead with most consultants, BBS pushes for a Zero At-Risk Behavior concept. This is a "leading indicator" approach and obviously an improved means of measuring safety in a workplace over a lagging indicator such as worker injury frequency.

There are various methods of incorporating BBS into a work

group but basically it begins with the identification of a list of "at risk" behaviors that, if used, will likely result in injury. Job site audits are used to gage the percentage of "at risk" behaviors versus "safe" behavior. The object, of course, is to continually increase the percentage of safe behaviors over at risk behaviors.

With BBS, when you begin to involve the workers, you are not just trying to lower injury rates. You are trying to do something specific that prevents injury; that is "upstream," if you will, of the occurrence of an injury.

The application of these processes are effective and have proven to be a valid methodology in altering the workers' willingness to use "at-risk" behavior in executing their work. Some of these BBS processes utilize the workers themselves in measuring their own "at-risk" behavior. Others use supervisors to do the observations. If you use one of these processes, you train your employees to measure a project customized list of "safe" and "at-risk" behaviors.

Then, by spreading the BBS process to greater numbers of employees and thus observations, one can, in time, reduce dramatically the number of instances where employees are observed using "at-risk" behavior in accomplishing their work assignments. These measured results are posted where the employees can actually see the progress they are making. Typical reporting would show a chart of "Percent Safe Work." Percentages of "safe work" ranging up to and above 95 percentile are desired. Some achieve percentages "safe work" exceeding 99%.

BBS is gaining in popularity. There are videos, training materials, and software available from several of these proponents to assist you in applying this technology to your workplace. In a recent Internet search for BBS I found 20 Internet Web sites that used the words in the material found by the search engine. So, from this little piece of exploration, one can see where the current attention is being focused.

The BBS approach is just what it advertises to be; that is a worker involvement tool. It is viable, good and powerful if used with the proper amount of training and utilization of the application processes and if BBS is completely supported by management. The key words in the preceding sentence are "completely supported." For this support to be institutionalized, the managers and supervisors will need training as well. While BBS techniques are viable, as with any other technique, they must be properly structured and applied for successful long term injury reduction to be the result. I like the BBS products and think the application to be a powerful worker involvement process.

Remember, worker focused Behavior Based Safety can be rendered invalid if management does not have its own safety thinking structured and its safety "commitment" well defined.

The 2002 Findings

The CII "Making Zero Accidents Happen" Taskforce also performed research in the process industries where the operating units are subjected to periodic maintenance. These short duration maintenance episodes (frequently including capital additions and expansions) are called "Unit Shutdowns," "Unit Turnarounds" or "Unit Outages." I will use the term "Shutdown" to discuss the research results.

The unique and challenging feature of a Shutdown is that it has extremely rapid build up of activities and thus employees. As opposed to a much slower build up in the average construction project, a Shutdown presents a host of challenges not found in the average longer-term construction project. Seven-day workweeks, with 10 to 12 hour shifts are common. Detailed pre-planning of all phases of the work is of paramount importance including. Pre-shutdown planning of the safety aspects of the work and the work process are of vital importance.

The Taskforce did the research on 44 Process industry Shutdowns in Petro-Chemical, Paper, Power and other Industrial facilities. Amazingly, 22 of these Shutdowns, were executed with Zero OSHA Recordable injuries. Also, of the 44, 38 were completed with Zero Lost Time injuries. It goes without saying, that these contractors are the best of the best with the Recordable Incident Rate averaging 0.70 for all 44 Shutdowns.

The Taskforce wanted to determine what, if any, specific worker management techniques were used by those contractors that achieved Zero Recordable injuries in executing these Shutdowns.

The key research areas were:
1. People resources
2. Planning
3. Scheduling
4. Contract formation strategy covers safety
5. Degree of support of CII's Zero Accidents research

People –
- Bringing workers in 2 weeks before or sooner
  o Those who did RIR = 0.22
  o Those who did not RIR = 0.58
- Workers brought in that were familiar with the work
  o Those who did RIR = 0.47
  o Those who did not RIR = 2.28

Planning –
- Software used to schedule work
  o Those who did RIR = 0.72
  o Those who did not RIR = 2.92

- Schedule unit used to plan work
  - Days RIR = 1.51
  - Shifts RIR = 0.68
  - Hours RIR = 0.45
- Days worked per week
  - Seven RIR = 0.96
  - Six RIR = 0.38
- Shutdown duration
  - Two to eight weeks RIR = 1.20
  - Less than two weeks RIR = 0.62
- Crew size
  - Over 12 workers RIR = 1.62
  - Seven or less RIR = 0.55

Combination of People and Planning approach
-Worker familiar and scheduled by hour
  - Not familiar/hours not used RIR = 1.75
  - Familiar or Hours RIR = 1.08
  - Familiar AND Hours RIR = 0.28

Combination Planning and Scheduling
- Duration and Days worked
  - Four weeks at 7 days RIR = 1.36
  - Less than two weeks and six days RIR = 0.38

Contract Strategy
- Contract Incentivized for Zero Injury
  - No incentive RIR = 1.73
  - Yes to incentive RIR = 0.71
  - Craft get Incentives RIR=even better than 0.71
  - All get incentive if RIR = 0.00
  - No one gets if there is a recordable
  - Incentives awarded on weekly performance

An interesting and powerful safety motto for all on the shutdown was given by a Taskforce member was as follows:

"If it is not safe I will not do it and will not let others do it."

Such a motto really speaks to the heart of a Zero Injury culture where "every man is a safetyman." All have a fearless devotion to seeing the work completed without an injury to anyone.

The Taskforce further stated that the shutdowns that performed at the Zero Recordable Rate were only those that used the above approach to managing the work along with the use of all nine of the Zero Accident techniques shown on page 71 above.

CHAPTER 11

AN EXAMINATION OF NEAR MISS REPORTING
AND
INTRODUCING
THE EMPLOYEE SAFETY IMPROVEMENT CARD

The CII Research Summary

Safety culture research conducted by The Construction Industry Institute (CII) on one hundred twenty two (122) construction projects revealed that there were twenty-four (24) zero injury safety culture features that when all were used effectively produced a zero injury (recordable) result.

Significant among these twenty-four features were 3 dealing with the maintenance and support of a "Near-miss" reporting process, the formal documentation of the Near-miss reporting and tracking system with emphasis on the workers being encouraged to report Near-misses.

Further the research found the project having the most Near-miss reported by the employees was also the safest project producing a zero recordable result.

The immediate question arises of "Why might this be true?"

This question and others are addressed in this book with the objective being to produce an in-depth work on the general subject of Near-misses culminating in a clear description of how committed leaders must lead in order to create a safety culture yielding zero injury through the empowerment of an effective Near-miss reporting process.

The 24 CII Techniques are:

1. The president/senior company management reviews safety reports generated by the projects.

2. Top management has involvement in "injury/ incident / near- miss/ accident" investigations.

3. Management and supervision are evaluated on safety performance.

4. The project safety representatives report directly to company senior management.

5. The company maintains a minimum of one safety representative to 50 workers. (see page ??? for clarification.)

6. The project has a site-specific safety plan.

7. Before each task, a task safety analysis / pre- task planning meeting is held with the foreman's crew.

8. Safety training is a line item in the project budget.

9. Every worker on the project attends a standard orientation training session.

10. The safety orientation training is formal.

11. Workers receive an average of at least four hours of safety training each month.

12. Superintendents and project managers attend mandatory safety training sessions.

13. All levels of management and supervision receive training in behavior based safety management.

14. A structured worker-to-worker safety observation program is maintained.

15. The company/project supports and maintains an effective near-miss reporting process.

16. A formal documented system exists to report near-misses.

17. Workers are encouraged to report near-misses.

18. Safety recognition / rewards are given to the workers at least monthly.

19. Family members are included in safety recognition dinners.

20. Management and supervision are evaluated on safety performance.

21. Subcontractors are required to submit project specific safety plans.

22. Sanctions are imposed when subcontractors do not comply with safety requirements.

23. Safety perception surveys (worker input) are conducted on the projects.

24. Off-site company personnel perform frequent audits/ assessments.

By category one can see safety training to be the most cited with 6 items (8, 9, 10, 11, 12, 13); management involvement 4 (1, 2, 3, 20); worker involvement, recognition 4 (14, 18, 19, 23); safety reps 2 (4, 5); safety planning 2 (6, 7); subcontractor management 2 (21. 22); auditing 1 (24). All are important with Near-miss being the subject of three of the 24 (15, 16, 17).

There is little published work addressing the subject of Near-misses and their importance in understanding their impact on the process of creating a safety culture where the occurrence of an injury is an extremely rare event.

Safety Metrics

By safety metric category that focuses on eliminating employee injury, Near-misses should be placed in with those metrics termed "leading indicators." First allow me to give my definition of a "leading indicator." Safety metric indicators fall into two broad categories; "lagging indicators" and "leading indicators."

A lagging indicator is that measurable event that can be used to evaluate safety performance that is specific to injury and as a category event an injury has already happened. They, injuries, are historical and counting them measures our failure (as long as one injury remains) in guarding the health of our employees. We convert them to "rates of failure." If our Total Recordable Incident Rate is 5 it states we are performing at a employee injury failure rate of 5 injuries per 100 employees per year.

There are two significant problems with measuring rates of failure as a safety performance indicator. One is when we eliminate our failures we no longer have an indicator we can refer to that allows us to see how well we are performing in safety. Two, in such cases when we find our failure rate at zero we no longer have a positive metric telling us we what we are doing (leading "injury preventive" indicators) that will allow us to continue to measure our progress in injury prevention activities that will sustain the injury failure rate at zero.

This need for a "positive pre-injury leading metric" requires safety leaders to look for, measure and calculate rates for incident prevention activities; i.e., "leading indicators."

By category "leading indicators" are the things we do that are in some cases required by law (mandatory) and in other cases are those things we choose to do (voluntarily) that are optional as far as the law is concerned but from experience have proven to be effective in injury prevention.

This definition leads one to easily see that all injury prevention activities then that are planned, conducted or carried out to improve safety can thus be measured and a LIR (leading indicator rate) rate can be calculated similar to the injury measuring lagging indicator TRIR. Note the word "indicator" as opposed to "incident." We can plan to implement "indicators", while "incidents" are unplanned.

All 24 CII zero injury techniques are examples of leading indicators. All we have to do is to numerically summarize the activities required to implement these 24 to get an LIR.

To illustrate we can use safety training; simply calculate how many employee training hours should be expended to meet the training requirement of the above CII techniques and compare the actual hours trained to the planned hours. Or as an alternative metric for training calculate for your work group the planned nominal safety training hour rate in terms of 100 employees (ie, 4 hours per employee per year would be a rate of 400) and compare how closely you come in percentage to meeting your planned rate in your work group. Reaching and exceeding 100%+ should be the objective.

A pure Near-miss though is an unplanned event; so how can that become a leading indicator? Answer: It cannot; however the"fact finding" available coming out of that failure can become one of your leading indicators!

More on the CII Near Miss Research

Three different CII Zero Accident Taskforce research reports reveal the direct positive effect Near-miss reporting can have on safety performance. These reports are –
1. Focus on Shutdowns, Turnarounds and Outages, CII Report 160A-1; July 2002

2. Safety Plus: Making Zero Accidents a Reality, CII Report 160-1; February 2003

3. The Owner's Role in Construction Safety, CII Report 190-1; March 2003

The Taskforce Findings

Report:
"Focus on Shutdowns, Turnarounds and Outages"
CII Report 160A-1

The safer projects investigating all Near-misses recorded an average TRIR of 0.60. Those projects investigating only "serious" near-miss recorded a less impressive average TRIR of 2.40.

Under "Conclusions" this report stated the *"Safer projects conducted thorough Near-miss investigations with involvement of various personnel."*

Report: "Safety Plus"
CII report 160-1

In the 160-1 report narrative it is stated; *"Documenting Near-miss was found to be among the more influential factors in reducing accidents/incidences....tracking them may indicate a true commitment to achieving good safety performance. Among companies that did track Near-miss, findings suggest that relative RIRs were lower for those companies that recorded more Near-miss. This may imply that the more vigilant a company is in tracking conditions associated with potential injury sources, the greater the prevention of actual injuries."*

The safer projects investigated over 50 Near-misses each with the average TRIR being 0.55. Those projects investigating less than 50 Near-misses recorded a less impressive average TRIR of 2.40.

Report:
"The Owner's Role in Construction Safety"
CII report 190-1

The report states "*...when the owner's safety representative checks the project Near-miss rate on a regular basis, the project tends to achieve better safety performance.*"

On those projects where the owner representative checked the project near-miss rate on a regular basis the TRIR was 1.22 while those owners who did not check the Near-miss rate logged a higher TRIR of 2.18.

In the summary section the report states "*...the key measurements employed by the owners with better safety performances are identified as:...Owners monitor Near-miss rates...*"

Comments on Near Misses

This author observes that while Near-misses are given considerable space in the three reports it is clearly stated in but one of the three reports (Safety Plus) that the best correlation to lower TRIR exists as the number of Near-miss reports increase. The other reports simply conclude better safety results when "all" Near-misses are investigated and when the owner representative routinely "checks" the Near-miss log.

There is a concern to be clarified here and that is one cannot be totally informed on the aspects of the CII research on Near-miss reporting until all three of the CII research reports coverage of Near-miss is read and understood. In the 2003 Annual Conference presentation of the "Safety Plus" report John Mathis mentioned that the single construction project that was using all the above listed CII Top 24 features of a zero injury safety culture also had something over 400 Near-misses reported during the course of the project. Seeking an "effectiveness cutoff point" the CII Task Force Chaired by John Mathis of Bechtel chose 50 Near-misses.

The History of Near-miss Attention

For decades safety terminology has included the terms near-miss, near-hit, close call, good catch. All are aimed to describe that event in the progress of the work where exposure to injury "ran out of control." Someone could have been injured but as good fortune would have it no injury occurred.

I use the words "good fortune" advisedly because it is my conviction that in the world of safety the only time "good fortune" or "luck" is involved is at the exact instant the safety execution of a work process goes out of control. Up until that point our planning of safety can have the upper hand and CII research proves unwanted events can be avoided.

However, at the "exact instant" an unwanted incident begins it is largely "good luck, good fortune, or bad fortune or bad luck" that takes over and as the event unfolds its effect/outcome on property or personnel is driven by circumstance that is beyond the control of the observer/s. If no damage or injury occurs we call such an incident and further label it a near-miss, near-hit, close-call or good catch. If damage occurs we call it an accident. If an injury also occurs we add the human medical treatment involvement as a "first-aid, recordable, or DART* case."

For decades it has been for some common practice in safety management to keep a count on the number of Near-misses occurring per year in a workforce. It is also widely known that careful examination and investigation of Near-misses provide valuable material for tool-box safety meeting topics.

So the CII research outcome was a strong validation of what most all safety leaders already intuitively knew but lacked hard data on.

* DART – DA = Days Away, R= Restricted duty, T= job Transfer.

Near-misses In Other Places

As an event, Near-misses occur in many and varied settings. There are Near-misses in the Medical field, the Aviation field, the Firefighting field, and in Automotive traffic to name just four. Truth is, Near-misses occur in all enterprises and activity situations involving humans.

All these Near-miss micro-environments can be generally classified into two distinct groups; "uncontrollable" environment and "controllable" environment.

Define please –

"Uncontrollable" is that environment where other humans are involved that you do not know, have no way of influencing and are typically total strangers; i.e, aviation (Near-miss in flight is an example). The Federal Aviation Administration (FAA) has "rules of flight" that are in place to categorize, count, and publicize in order to minimize these unwanted Near-miss events. Automotive traffic is a similar example. Here the various states have traffic laws in place to minimize unwanted incidents. Yet you as an independent driver have no relationship, direct communication or influence on how another driver adheres to or ignores these rules. Such can be described as an "uncontrollable" environment and Near-misses can occur due to the random unexpected act of other drivers. "Defensive Driving" is a popular course designed to attempt to reduce accidents in driving situations by teaching you how to anticipate the unexpected; still the other driver is outside your sphere of influence or your control.

On the other-hand the profession of "firefighting" is an example of an occupation with both "controllable and uncontrollable" conditions; sometimes unknown (non-human) conditions or building structural situations that surround the fire cause incidents or Near-misses. Near-misses can also occur when the unknown condition asserts itself and someone is involved in a close-call. Yet the firefighters are typically a coordinated team and all know what to expect of a fellow firefighter and thus near-misses are minimized. In the latter, human situational firefighting training and planning reduces the Near-misses, but even to this degree are still only partially "controllable."

In the medical field for instance an operating room active with a surgery underway is a very "controlled" environment. The medical team is highly trained and accustomed to working together. All this serves to minimize Near-misses during the surgical procedure. Even with occasional medical equipment breakdown, planning has set aside backup equipment to keep the patient safe from harm in that event. It is a "controllable risk environment" and Near-misses can be and are minimized.

All the above discussion is given to point out that most work-places can be "controlled environments" if the leaders put the appropriate energy into making it so.

Safety professionals are engaged to ensure that all applicable safety activities are made known to the leaders in order that they might incorporate them into the daily work planning.

For instance, in a construction project environment safety technology is applied by the leaders with the aid of the safety professionals to eliminate at-risk situations along with the use of unsafe (at-risk) behavior and thus minimize safety incidents including Near-misses. With all employees being part of the safety team and all given the same safety orientation and training the environment becomes "controllable."

The better the training and orientation and the more the employee "buys-in" fewer safety incidences and Near-misses will be experienced. In a tightly controlled construction safety environment even visitors to the site are put through safety orientation in order to further "control" the environment.

Employee buy-in

Obviously, despite the leader's efforts to "control the environment" it ultimately is each employee's choice as to whether or not they will "buy-in" to the safety culture and effectively participate. The population of employees that "buy-in" is directly proportional to how effective the project leadership is in informing, educating and persuading each employee to become an active part of the safety culture.

With this reality the question then becomes "How do I as a leader go about successfully persuading each employee to become an active cooperating part of the safety culture?"

The answer lies in how one applies a battery of leadership skills designed to accomplish this. When we see projects achieve remarkable safety records we look closely and copy their leadership processes.

Near-miss Reporting Policy

It seems all companies engaged in the construction industry have some kind of Near-miss reporting policy even if it is undocumented and informal. Even with a near-miss reporting policy in place those company and project leaders I have talked to regarding Near-miss reporting without exception found getting the crafts to report Near-misses to be a very difficult task.

Again we have the same question arise of "Why might this be true?"

My aim is to address these questions and seek answers that are usable by those wishing to understand the Near-miss reporting effect on safety performance and the reluctance of the workforce to report Near-miss. More importantly my purpose is to explore in-depth the culture issues that prevent Near-miss reporting and provide insight to the leadership processes that have proven to maximize worker Near-miss reporting.

How many Near-misses are there?

In my consulting career I have interviewed hundreds of crafts personnel in traditional safety cultures with questions regarding the job site's safety culture as they perceive it.

In addition from time to time I have asked selected crafts personnel (working in a traditional approach to safety leadership; non-zero injury focused) to estimate from their experience how many Near-misses would occur in a crew of 10 craft workers over the course of a year. To do this I defined a Near-miss as "any event that could have resulted in even the slightest injury, property damage or worse; but did not."

The estimates given out of traditional safety cultures have ranged widely from one Near-miss per quarter per worker to one per month per worker. Transferring these estimates to the average "hours worked base" for 100 workers per year we get a range from 400 to 1200 Near-misses per year per

200,000 hours worked. If you wanted to reduce this numbers to a Near-miss Incident Rate the rate would be 400 to 1200!

On a national scale we know the "Near-miss like" events that result in injury translate to the 2009 OSHA TRIR average of 4.3 Using the lower 400 estimate and dividing by 4.3 we see that for each Recordable injury reported there would be 93 Near-misses; a 93 to one ratio. Honestly no one knows far a fact how many Near-misses occur but these numbers (93 to 1 ratio) or 93x4.3= represent the lower average from those crafts-persons I have asked.

Adding the subjective Information

In the CII Research Report "The Owner's Role in Construction Safety" page 4 gives an injury pyramid showing the ratio of Lost-time injuries to OSHA Recordable injuries, and the ratio of OSHA Recordable injuries to First-aid injuries. Here I am taking the liberty to add my subjective information from the interviews with the crafts and add a ratio for number of Near-miss per First-aids.

Pyramid sources.

I find in an internet search that Frank Bird, of Loganville, Georgia has written a text book that includes a ration pyramid. He suggests a ratio of 1 -10 – 30 - 600 defined as a progression from 1 major injury to 10 minor injuries to 30 equipment damage events to 600 Near-miss.

The numbers are expressed as Incident Count or Rates.

My number gained from subjective opinion research (93 per recordable or 400 per 100 employees) if added into a pyramid is not intended to reflect hard data but to remind the reader that there is huge number of Near-misses occurring in the workplace and we are truly not at all informed on the ratio so perhaps a ratio derived from subjective data is better than none at all.

The relative (400 vs 600) close agreement with Frank Bird's number of 600 is purely coincidental for I was not aware of his ratios until some days after I made my calculations. Factually speaking we know that the actual reports by the workforce (low) numbers are not at all accurate to even attempt to calculate a ratio. Such would be far too low. For Bird's triangle see the next page.

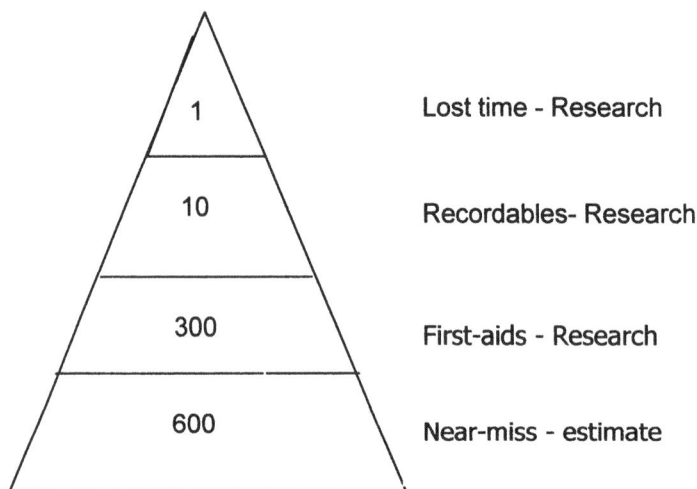

1	Lost time - Research
10	Recordables- Research
300	First-aids - Research
600	Near-miss - estimate

Figure 1

Reworking the above triangle math to indicate number of Near-misses per recordable yields the following relationships:

One recordable per 60 Near-misses.

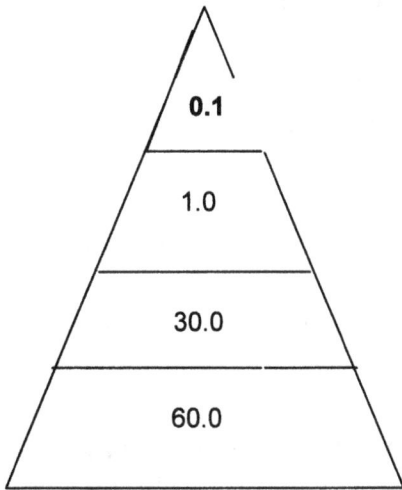

```
        /\
       /  \
      / 0.1 \
     /------\
    /   1.0   \
   /----------\
  /    30.0     \
 /--------------\
/      60.0       \
/------------------\
```

Figure 2

-

This is just a reminder; Near-miss reporting is very important as a Leading Indicator! Another way to view the "Near-miss to Recordable" ratio is as a safety leader sets in motion safety initiatives to reduce Near-miss; i.e., identifying, reporting and investigating Near-misses that this whole process centers around "safety awareness" on the part of the workforce.

This awareness is what creates a lower "at-risk" behavior rate hence lower injury rate. So it is natural to conclude as your recordable rate drops below the 2009 national average (in construction) of 4.3 to say 3.00 or even 1.0 this is occurring because "safety planning, awareness and safety rule adherence" has effectively reduced the number of "at-risk" behaviors hence the number of "near-misses" that are occurring.

Projects Reporting the most Near-miss were the Safest.

Why might this be? There are at least 3 reasons;

1. Leaders clearly defining a Near-miss for the workforce provide a clear understanding of what to report thus more accurate reporting.

2. Applied energy by employees seeking to identify and report Near-miss maintains a safety focus in the minds and conversation of the workers as they go about their workday; and finally

3. Near-miss investigations reveal weaknesses in the work processes that allow safety exposure while the identification and correction of these weaknesses brings strength to the incident prevention side of safety.

All three work together resulting in fewer near-misses hence fewer recordables.

Near Miss Reporting

In order to persuade a workforce to report all Near-miss leaders must successfully remove the stigma that naturally prevent full Near-miss reporting. These are "fear of retribution" and "fear of ridicule." Foundational to success in removing these stigma is to publish a clear definition of a Near-miss in simple terms. Creating this definition is more complicated than it appears.

Defining clearly to the workforce just what constitutes a near-miss is a critical element in creating a culture where employees freely report near-misses. In my view the often used simple definition of a near-miss is "Any event that could have caused injury or damage but did not" while true, is an inadequate definition in and of itself.

Why? It needs further clarification and some examples for the leaders and workers to model reporting after. One caution here; be very liberal and allow all "if an injury could have occurred and did not" events to count. So please do not get real picky.

Considerations in Defining a Near-miss

For instance do you wish to include the violation of a safety rule or procedure as a Near-miss; such as if someone is seen without a hard-hat in a mandatory hard-hat area is this to be termed a near-miss; or is a ladder in use and not tied off in the prescribed manner to be registered as a near miss? Or are you going to restrict the definition of a near-miss only to those events similar to when some object falls near an employee without a hard-hat; if such occurs then how close to the employee does the object have to fall, or does the object actually have to hit the employee, which makes the event an incident? Or does the untied ladder have to slip and the fall with a worker on it that is uninjured for it to be called a Near-miss? In these two examples no injury occurred and no damage was done fitting the definition of a near-miss.

I am not too sure as I think about it that good near-miss reporting will ever be possible because of the "safety rule violation issue."

Some would say while the breaking of a safety rule can result in a near-injury such should never be classified as a near-miss; when in truth it was a violation of a known safety requirement that allowed the near injury to occur in the first place. After all, the violated safety rule was in place to prevent the exact thing that occurred.

Those espousing this school of thought say these events are more simply called "safety violations" and should never be classified as near-misses even if the circumstances produced a hazardous event when someone could have been injured but was not. These events are identifiable and preventable because the lessons have already been learned and safety interventions have been designed to avoid such events.

It is the function of safety training to inform all employees of the safety rules and procedures in order that a worksite can be declared "safe." A "safe" worksite can only exist when all employees have been given the prescribed safety orientation and training modules and testing has been accomplished to ensure understanding. Any group of workers or even a single worker not yet safety trained should create an anxiety in the hearts and mind of the project leaders that one of the untrained can be injured at any time due to lack of specific knowledge. If leaders allow an on-site worker "at the tools" it surely can be viewed as "at-risk" behavior.

Include Safety Rule Violation in a near-miss count?

Others would say safety rule violations must be included as near-miss because the problem is that many near-misses occur to individual workers when there are no witnesses. Some feel large numbers of these events occur as a result of inadequate attention or adherence to published safety rules and these events point to the need for clear near-miss definitions.

For example, a 2x4 is left by a trade upon an elevated platform where the wind can dislodge it or perhaps an electrical cord is left to form a tripping hazard to passers by. In defining a near-miss does the 2x4 have to fall near someone for the near-miss to be declared or can the near-miss be declared if someone discovers the 2x4 that could fall.

Or does someone have to trip over the electrical cord and not be injured for the near-miss to be declared or can a near-miss be declared by someone simply discovering the tripping hazard. Yet in both these illustrations safety rules are in place warning against each of these practices; thus each is a clear violation of safe job site policy.

The question then is, if a safety rule violation is not to be defined as a near-miss then does one need to establish two types of near-miss? 1. rule violation near-miss and 2. The other type near-miss?

It is obvious that employees are very reluctant to report violations of safety rules out of fear and or peer pressure plus it is not logical to assume employees will report safety violations on themselves or each other except in cultures where leaders have clearly proven to the workforce that such reports are welcome, even rewarded and no punishment is forthcoming.

My thought is to define rule violations as such and take that non-compliance out of the definition of a near-miss.

Reluctant Workers

The reporting, investigation and feedback to employees of lessons learned is one of the critical elements of creating a successful zero injury safety culture. If workers are informed of this in the safety training and intuitively agree then why are workers so reluctant to report near-miss?

On a number of occasions I have asked selected journeymen (I typically ask those who reflect an acute interest in the subject) during my safety culture assessments why workers are so reluctant to report near-misses. The answers received are not surprising and are intuitively known by most if not all construction project leaders.

The reasons are: a. fear of retribution, b. feeling embarrassment, c. maintaining peer relationships and d. it complicates my life and e. I do not want to take the time involved.

The Effect of Punishment

A near-miss can be defined as:
> **"an unexpected event or sequence of events that created unanticipated exposure conditions where except for 'good fortune' an employee was not injured nor damage done."**

When we add the choice of punishment, appropriately I feel, for safety rule and/or safe procedure violations we get the fear factors mentioned above and in this we see the basic cause of the virtual dearth of Near-miss reports prevalent on most jobs. This fear is driven by the unknown; the employee fears because s/he does not know but what someone reviewing the event will declare the employee has violated a safety rule. Thus reports are not forthcoming.

I find that most workplaces I have encountered are not effectively encouraging the reporting of near-misses even though they know the value and are actively asking that near-misses be reported.

Eggshells to Pavement

When one is trying to create open and free (as in fearless) reporting of near-misses one should know they are figuratively walking on eggshells. Once created such a culture is very, very, very vulnerable (one could even add more 'verys) to fracturing. This is why Near-miss reporting has not been successful but in rare cases. Even wwith all processes working in perfection, without even knowing it leaders blunder in the handling of the investigations and reports stop flowing. After this all reports come from those supervisors witnessed or become common knowledge for another reason..

The next question is; "How can we turn this 'eggshells' walking surface into a solid surface so these events do get reported."

Answer; "Make the reporting of s Mear-miss event into a totally positive experience with no fear of punishment."

How? Re-create the entire process of reporting!

Use a "Employee Safety Improvement Card" (ESIC).

Why, what, how, when?

Why? Properly designed and utilized it is a non-threatening tool to allow employees to highlight a safety issue, idea or Near-miss without the threat or fear of pending punishment from yet unknown sources for unknown reasons.

What? The card is a safety improvement recommendation report document that is "employee participation" designed 5.5 x 8.5 two sided card stock form. Print how the card is to be used on one side and "safety idea tickers" the other with space to record the details of "a good safety idea whose time has come." The few ticklers aan be culture specific and simple to understand and fill-in.

How? With employee participation in the creation of the form the form immediately becomes employee owned. Give them the reins and let them have some of the creation process. This way you not only get great safety involvement but you get those inputs from the workforce that once acted upon creates the safety culture ownership you so critically need for a zero at-risk behavior safety culture. Ensure in the process that check slots are included that ask for ideas out of any Near-miss experienced by employees. Do not make it mandatory that an idea has to have an idea originator. Allow the ideas to flow anonymously from any and all. Signing the form is optional. Publish a list weekly stating what was done with the ideas turned in. Make the commentary very complimentary and supportive. It is ok to state the idea is under consideration but never let an idea just drop.

The ESI Card has tremendous power for culture building.
The ESI Card straightens the process of reporting, levels the mountains of uncertainty and fear, and paves and smoothes overall safety improvement communications.

All is done while ensuring no idea is ever criticized but welcomed with open minds and hands as the culture leaders strive to find ways to implement even the smallest notion.

Remember this:
A zero injury safety culture is best created as many employee created seemingly small ideas are applied allowing culture ownership and diminished at-risk behavior to occur with credit given first of all to those with the highest exposure that worked injury free.

When? The ESI Card is not a new idea but has been tried with success in a few companies. The time to begin use is now!

Obviously the use of expertly leader performed "no-fault" idea explorations (rather than investigations) will be the required norm and is even mandatory. The term "expertly done" is not mentioned lightly for the slightest slipup in language use during the exploration that an offering employee can interpret as judgmental or condemning (the rationale behind the offending words notwithstanding) will get back to the workforce and will serve to "shutdown" open and fearless reporting.

A single use of punishment arising out of the ESI Card use in such cultures will have a major negative impact on inclination to report. Therefore it goes without saying that effective evaluation training in process and language use during "no-fault" evaluations will be of prime importance. So much so, as to prompt the following observation of -

> **"To not specifically train leaders for no-fault explorations is to guarantee failure in creating a reporting safety culture where Near-misses finally find their way into your safety improvement processes."**

Remember punishment is the traditional approach. You have to change to a mercy based (no-fault) approach and such is a natural part of the CII "employee involvement" feature of a zero injury culture of safety.

My conclusion is the advertent or the inadvertent choice of punishment will cause reporting to go unreported.

So the choice is punishment or mercy. For a safety idea to be reported the only choice is "mercy."

Reporting of near-misses via the ESI Card is an essential ingredient to sustaining a zero injury outcome. Pure Near-miss reporting may well be a secret so hard to capture that only a rare few projects ever accomplish the objective but you can find success with a new non-threatening approach.

See next page for an example ESIC.

Employee Safety Improvement Card

This card was created by craft employees on this project. It provides a way to improve your safety by harvesting your ideas.

Complete the reverse side of the card and hand it to a Safety Specialist or a member of the Craft Safety Committee.

Your idea will be reviewed by the project Craft Safety Committee and feedback posted in the weekly Craft Safety Idea Review News Letter.

This card is numbered so make a note of the number and look it up in the Craft Safety Idea Review Review Newsletter that is published weekly.

Due to the number of ideas being generated it will appear only one time. If you miss it you can ask your Safety Specialist for a printout of the decision on the idea by using the number on the form.

Employee Safety Improvement Card - # 201

_____
_____
_____
_____
_____
_____
_____
_____
_____
_____

Source of Idea:
Past safety experience___   Observed Near-miss this project___
Observation on this project___   Sign if you wish:_____
Please indicate your craft:_____

Chapter 12

Demonstrating
Caring
A Most Critical Element

Caring

During the past 18 years The Construction Industry Institute (CII) has performed five batteries of research to answer the question; "Why can some contractors work millions of hours without a worker OSHA recordable injury while most cannot?" The reports of these research teams surfaced 24 safety interventions that were being used where safety performance at the zero recordable level was being experienced.

First and foremost (Number 1) was the absolute necessity that "Leaders Demonstrate their Safety Commitment." This leader commitment is revealed to the employees through the active support and use of the other 23 CII zero injury techniques. The first CII Zero Injury Task Force found, not surprisingly, that in applying the techniques "Quality of Effort" had equal standing with "Implementation."

**Creating a Zero Injury Culture of Safety
Requires Demonstrated Caring and
Treating Employees with Dignity and Respect**

After "properly" embracing the 24 techniques, practitioners find that incident rates plummet to near zero and records of one million hours recordable free are common. Note the emphasis on "properly" for I will discuss this aspect more below in this article. Data from the CII membership in 2009 reveal a TRIR of 0.63 and a DART IR of 0.16 for with 2.3 billion work hours reported; this, while the US OSHA/BLS average for 2009 was a TRIR of 4.3 and a DART IR of 2.3.

The immediate question is; "Why such sterling safety performance?" Strong evidence points to the "worker involvement processes" found in the 24 Top Techniques. The obvious reason "involvement" works is that when workers through "involvement" are made co-creators of the safety culture they naturally become more supportive of the safe culture because they are actively involved in creating it.

To bring focus on the importance of "involvement;" Did you know that worker "involvement" findings make up nine of the Top 24 CII Techniques? Nine of twenty four is 37%. This 37% are processes that "involve/encourage and recognize" the workers as partners in the design and execution of the safety program elements.

The nine are: a. worker to worker safety observation, b. an effective near miss reporting process, c. a formal system to report near misses, d. means to encourage workers to report near misses, e. monthly worker recognition while zero is being achieved, f. family members are included in safety celebrations, g. workers (and foremen) are evaluated (mostly praised) on individual safety performance, and h. worker safety perception surveys are conducted, i. incident investigations including near misses.

The Employee's "Wants" Matter

Here is what employees say they want, starting with what's most important to them:
http://www.selfgrowth.com/articles/Dunn110.html

1. Full appreciation for work done
2. Feeling "in" on things
3. Sympathetic help on personal problems
4. Job security
5. Good wages
6. Interesting work
7. Promotion/growth opportunities
8. Personal loyalty to workers
9. Good working conditions
10. Tactful discipline

Now take a look at what managers THINK employees want, starting with what they think is most important:

1. Good wages
2. Job security
3. Promotion/growth opportunities
4. Good working conditions
5. Interesting work
6. Personal loyalty to workers
7. Tactful discipline
8. Full appreciation for work done
9. Sympathetic help with personal problems
10. Feeling "in" on things

From this information you can see why we as leaders get it all backwards. It does help in understanding this if in your past you have been one of these workers.

Seeing that "worker involvement" in creating a safety culture is so critical, a question arises: "What can we do to ensure the foundational elements of a "caring, communicating and committed" safety culture are in place?

We can call them the three "C's."

For leaders to know the nature of people as a fundamental skill set is self evident. Some leaders are gifted with these fundamentals, others must be taught. To not train your leaders in the fundamentals of leadership from a safety perspective could be viewed as at-risk behavior.

Safety leadership style is critical. Can we define the correct style?

To find the answer lets' use the "root cause concept."

Root Cause Analysis

"Root Cause," analysis has been touted for a decade or more as one of the most effective processes to use in the investigation of safety incidents, accidents, injuries and near misses. It is a tool that allows the investigating team to "drill down" into the "nominal causes" seeking the "real cause" that lies at the root of the sequence of events ending in the safety incident.

The use of this tool in "drilling down" raises an interesting question.

Could one reverse the "incident investigation" aspects of a "Root Cause analysis" and apply it to determine why, in companies that "properly" use the CII 24, there is a virtual absence of unwanted safety incidents. Some are calling such work cultures by the name of "Zero Injury?"

Let's assume an investigator decides to use the "Root Cause" process in reverse, to "drill up" into the safety culture? To "drill up" from the very basement of a zero injury culture, the worker, seeking the safety culture characteristics that are the "root cause" of an injury free workplace; in other words, "What culture characteristics are found and what will be determined to be the "root cause" of the incident/injury free workplace?"

Is it not the absence of "at-risk behavior!"

What then is the leadership characteristics used in a culture that is incident free?

Using the root cause approach: since incidences are largely caused by at-risk behavior, what causes at-risk behavior?

Is it not "at-risk thinking?"

If it is "at-risk thinking" what is the root cause of this thinking?

Three things come to mind-
    1. Lack of knowledge
    2. Lack of buy-in
    3. Not thinking!

The Question then becomes; "How does a leader conquer these three?"

Is it not through caring, communication and commitment?

Caring:

One critical culture feature found is that when it comes to productive safety cultures the most effective cultures refrain from the use of employee harassment through the use of "threats, fear and arbitrary punishment." This is not to say that each of the 24 techniques is not important, but it is saying that even if you use them all and advertently or inadvertently do things that alienate your workers you are losing their essential morale and co-ownership/buy-in to the safety culture you are trying to create.

Personal field experience in applying the above listed eight involvement techniques has surfaced the absolute necessity of altering the traditional element in the mind of the workers of "fear of punishment" to one of respect for the zero at-risk emphasis as the leaders prove daily they sincerely care about the worker's well being and thus use a caring approach as the primary means of safety rule enforcement.

It seems logical, if leaders truly demonstrate they wish to invite workers to be an integral part of the safety team, then the workers must be treated as full team members by those team leaders. The routine use of criticism, threats and disrespect to push safety emphasis are out. The key phrase is to "Win the minds and hearts of the workers" to the point they feel as (because they actually are) co-owners of the safety culture. You do this with frequent leader expressed appreciation and praise.

Since we have placed "at-risk" behavior as the cause of most all safety incidents it rapidly becomes obvious that for a zero injury culture to exist "at-risk" behavior must be eliminated. You might disagree at first but please allow me this statement; "Most employee 'at-risk' behavior is caused by or allowed to occur by ignoring the basic CII technique premises of "appropriate involvement" which results in "winning the mind of the worker" Despite concentrated effort by some leaders often not all leaders are "bought-into" the "caring approach" thus some leaders remain traditional, uninformed or simply inept.

If true, why might this be so?

The reasons are as follows and are found from the beginning of the workers' employment in the hiring process through the early weeks of employment by "leaders" failing to take advantage of the naturally occurring opportunities to instill a "free of fear" sense of "safety self-accountability" into each employee hired.

Developing Good Safety Leadership
The creation of a zero injury safety culture is largely dependent on "leadership" because of the following:
1. Leaders set the tone of employment; "attract or alienate."
2. Leaders set safety expectations in the hiring process.
3. Leaders provide the safety training facility and its ambiance.
4. Leaders provide the "quality" of the safety training.
5. Leaders provide the safety education and training curriculum.
6. Leaders provide site specific safety orientation.
7. Leaders provide safe tools.
8. Leaders provide safety planning processes.
9. Leaders provide safety coaching to the workers.
10. Leaders provide opportunities for craft involvement in safety.
11. Leaders provide the "employee to employee" relationship model.
12. Leaders instill a "sense of being respected."
13. Leaders are the safety role models to the workers.
14. Leaders are the lead safety communicators.
15. Leaders are the lead investigators of incidents.

16. Leaders share the results of investigations.

17. Leaders choose their leadership style.

18. Leaders choose "no fault" or "fear."

Quickly one can see that what is brought to the job site by the worker is "mind and muscle." It is the task of Leaders to welcome the new worker in such a manner as to promote their feelings of being wanted and appreciated. It is the job of Leaders to envelop the new worker into the safety culture through caring, communication and the demonstration of Leader commitment to the absolute safety of the "worker."

Too many leaders spend all their time focusing on the muscle; when in fact they should be spending quality time "winning the worker's mind" in safety.

With the above leader roles described one quickly realizes that Leader Training is extremely essential. This item too is one of the CII Top 24. How many employers actually train leaders in "how to Care?"

In summary to get to absolute zero for longer and longer periods of time we must once again embrace something different and in my experience the very best performers focus heavily on maintaining a cadre of leaders who are extra-ordinarily skilled in interpersonal relations.

A final statement:
> *It has been proven many times over in the past decade that a company can install all the 24 Top CII Techniques and still fail to create a zero injury outcome.*

Why might this be? The reason has been found in the inadvertent alienation of the employee through inconsiderate supervisory techniques that prove to be punishment oriented.

Why do we find this "punitive approach" flourishing? I believe it is found in our traditions. Prior to the advent of the Zero Injury Safety Leadership Concept being discovered most employers either did little in establishing safety process and procedure or used a "my way or the highway" approach. Back in 1989 In the "heyday" of traditional safety leadership the BLS/OSHA national average TRIR for construction safety was 14.4. In 2009 it was 4.3.

Why the improvement?

There are many reasons and one of them was the advent of the Zero Injury approach to safety!

Research found the contractors achieving the best results were involving their employees in the nine ways listed in the Top 24 CII Techniques.

"Winning the mind of the worker" can never be properly and successfully accomplished through threats, intimidation and punishment.

But, "mind-winning" can be accomplished through respect, kindness and careful communications that focus on involving the worker in the job culture creating process.

One of these means of showing respect and kindness is found in item "15." in the above list. "Careful" investigation into the causes of any safety incident extending even to near misses is a good example of the potential for showing caring, kindness and worker involvement.

It is important to design these investigations to involve the workers in the process, again "in non-threatening ways" without the "fear" component. Some call these "no-fault" investigations, where the facts are sought while assuring the injured and witnesses that punishment is not forthcoming as a result of the investigation. But safety training re-training will likely be.

Right here some will say, "But what if fault is found, a safety rule violation was discovered? Should not the violators be punished?" The answer is found in the use of "Mercy." Many, many times it is the recognized presence of demonstrated "Mercy" that allows the incident to be reported in the first place. If the "fear of punishment" is allowed to be present then incidents will go unreported. It is human nature to protect one's self from punishment. When Leaders understand that the safest workplace is found where "Mercy resides," then true success in winning the "minds and hearts" of the workers is realized and a true Zero Injury culture can come into being.

"But, hold up here a minute," you say, "what do I do about that individual that insists on engaging in "at-risk" behavior, even after being warned by their supervision, often more than once?

This is where the concept of "self termination" comes in. Simply put, as a leader you set safety performance expectations. As these expectations are presented be very careful to do so in a kind and caring way that avoids threats and harassing remarks, such as "You do this and you are gone!" Rather use the logic of "caring."

An example would be found in the following leader's remark "Please do not engage in 'at-risk' behavior of this type for you see if I/we allow you to place yourself at-risk then I/we are as much to blame if you are injured as you are. Since I/we care about your well being we feel it is unfair for you to place us in this position so please avoid all at-risk behavior in the future.

"You see, we believe, that willful safety violations (willful at-risk behavior) cannot co-exist with our zero injury safety culture. The question becomes then, How do we deal with willful violators such as you? The answer is we will use "Mercy.""

Sometimes the violations will indicate the need for additional safety training. In such cases you train first. Then you counsel the violator. To illustrate the leader/employee conversation allow me to give the violator a name. We will name the violator "Ur Badhabit." The conversation will go like this:

Leader: Ur, I would like to talk to you about your safety performance regarding your PPE. You have been reported violating a known safety rule. You need to know that as your leader I am trying to protect you and your co-workers from being harmed on this job.

To accomplish this we need everyone to abide by the safety rules for three reasons: 1. Not properly wearing your PPE can lead to your injury; 2. by violating a safety rule you are being a poor example to your co-workers, and 3. by me allowing you to violate a safety rule I am condoning your conduct and you are challenging my role as your leader.

If I allow you to violate a safety rule then I become an accomplice in your act. This I will not do. So at this time I am going to allow you to decide whether or not you are going to continue to work here.

Ur, if you choose to abide by our safety rules, and I sincerely hope you will, then you choose to continue to work; however, if you choose to violate a safety rule again, then by that very act you are saying to me, that you are choosing to terminate your employment.

As of now, it is your choice.

As I said, I hope you choose the safe way. I care about you. We need your skills.

I urge you to work safe!

This approach to safety compliance must be explained in detail at the time of employment and reinforced by the immediate supervisor when the employee reports to their supervisor.

This approach is a caring, considerate action that communicates leader safety commitment in certain terms that uses the concept of Mercy and Self-termination.

The typical worker reaction is – you have a newly committed safe employee on your job.

And the TRIR will continue to fall.

Think about it!

CHAPTER 13

THE SAFETY QUESTION OF THE 21ST CENTURY
"ARE YOU READY YET?"

Zero Injury is Here

Fact: Change is a necessary element of progress in any field.

> One could say that positive change,
> straightens, levels and paves,
> the highways of progress.

Positive safety progress, as measured by BLS/OSHA incidence rates, in reducing injury in the American work place has been substantial over the past decade. In 1989 the BLS/OSHA construction industry injury recordable rate was at 14.3 per 100 workers per year; in 2009 the rate was 4.3, an improvement of 70%. The root of this improvement has been a willingness to change how employee safety is managed on the part of all the many contributors to this progress.

The BLS/OSHA rates cited above are known by those knowledgeable in safety matters as "lagging indicators." These numbers as they are calculated are simply failure rates. These rates tell us after the injuries occurred what the rate was, a historical injury figure as a percentage of 100

128

workers on an annual basis. This "lagging" measurement process has been used by OSHA to measure safety intervention effectiveness since the early 1970's when OSHA was established.

Question: Are the leaders of industry and the safety professions ready to make an operational change from the "lagging indicator;" i.e., BLS/OSHA injury rate measures, to a "leading indicator accident prevention predictor" model.

It is now well known that there is a publically available battery of injury prevention research based knowledge that if used effectively does in fact predict a zero injury outcome.

Here the disbelievers cite the law of probability and say "hog-wash, zero injury is not probable." I completely agree if one takes a "zero forever" perspective. Sure it is obvious, given the nature of man that to deduce "a zero forever" conclusion founded on pure logic; can't happen. And I agree totally.

HOWEVER, if one takes the shorter view; one where we are trying to work injury free today then it becomes not only possible but is happening everyday in all companies.

This simple fact highlights the following;

"The probability of zero injury for the long term not being statistically probable,
does not remove
the possibility of zero injury for the short term being statistically possible!"

Let's take a look at BLS/OSHA national averages as an example. The 2008 BLS national average for all industry was a "failure rate" of 4.2 Recordables per 100 workers per year. Turning this number around into a "success rate" reveals that 100 workers worked a total of 245.8 days of a

250 workday year recordable injury free. To get to whole numbers for every 500 employees 22 (4.2x5=21) are injured. But as a percentage of time using an 8 hour day norm, industry worked recordable free 98.32% (245.8/250=.9832) of the work-days in 2008. Viewed in this way things look a lot better, except of course if viewed by those 21 of the 500 that were injured the 1.68% below 100% does not look so good at all.

Common practice today is to measure "failure rate" (the 4.2%) and most all of the time the principle remedy is to examine the failures to learn how to prevent similar accidents in the future. As this is done one is in fact "behind the power curve" in measuring true incident prevention activity.

In 1970 OSHA was established in an effort to "head off" the increasing number of accidents and injuries and set about to stop them by fostering reduction of exposure through interventions in the work environment. The agency itself embraces the concept of "accident intervention" by setting safe standards of operation to protect the many occupations present in American commercial enterprise. But typically OSHA works "behind the power curve" as well, using accident data, at least in part, to establish the next rule making focus area to reduce the lagging indicator of injury/ incident/accident rates even further. Informed people believe that some success has come from the OSHA approach.

But another approach was taken by the Airline Industry. An entire world wide industry employing nearly one million people was born in the 20th Century when air travel became common place. It is currently reported that there are over 2000 airlines operating worldwide. Today, the global air transport industry operates more than 23,000 aircraft, providing service to over 3700 airports. In 2006, the world's airlines flew almost 28 million scheduled flight departures and carried over 2 billion passengers with safety as a major concern. The following is taken from the New York Times.

October 1, 2007

> WASHINGTON, Sept. 30 — After two infamous crashes in 1996 that together killed 375 people, a White House commission told the airline industry and its regulators to reduce the domestic rate of fatal accidents there has been strong progress internationally. William R. Voss, president of the Flight Safety Foundation, recently calculated that if the 1996 accident rate had remained the same in 2006, there would have been 30 major accidents last year. Instead, there were 11.0 percent over 10 years. That clock ended Sunday. They have come close to reaching that goal. Barring a crash before midnight Sunday, the drop in the accident rate will be about 65 percent, to one fatal accident in about 4.5 million departures, from one in nearly 2 million in 1997.

"It's not one thing. It's a series of 'small things'," said John Cox, who was an Air Line Pilots Association safety representative for 20 years. Many of those small things were minor problems observed in everyday operations, he said, then counted, scrutinized and eliminated before they caused an accident. This can be called working "ahead of the power curve."

The same approach can be used in any industry to reduce employee injuries. The research performed in the construction industry can be an example of this. After looking at and performing detailed analysis of 122 projects with hundreds of contractors the researchers surfaced nine major categories of leadership effort that was producing the very best in safety performance, with some results being zero OSHA recordables for a million hours and more.

Within these nine categories ZeroInjuryInstitute.com have identified 131 "small things" that were "leading indicators" in use by those contractors either at or closest to a zero injury result. The "small things" are predominately of a type one could call "people focused technologies." The term "technologies" is used to bring a proper focus to these "small things." Each has within its' successful implementation details that when closely examined are but small technical, people focused, details of how to go about a proper implementation of the technique. Sometimes these details are to be carried out by line employees, other times by line managers, and sometimes by the safety staff and not a few times by top company leadership.

There arises a common question when one learns of the research results out of the construction industry. It is: "How can safety research in construction be applied in my business when it is so different?" The answer is found in the fact that the 131 safety techniques are all aimed at how leaders go about influencing their employees to avoid at-risk behavior. In short the items are people focused safety strategies you use to incorporate your employees in the safety mission. If your safety mission is not to avoid "all injury" then what is it. If it is less than zero you are thereby saying it is OK for some of my employees to be injured! Incorporating the 131 leading indicator techniques into your safety program takes you out of managing a failure rate into managing your success rate. Your safety audits seek to find how well you are doing in implementing the leading indicators.

Within the fine tuning of the functioning of these 131 "small things" leading indicators, there was found, what one could call the ultimate in success; zero Injury.

Sadly we see many think by simply "demanding a zero injury result" they can find success. This tactic has been found to fail in all cases. Conversely many have found true success by studiously implementing the 131 leading indicators. It sounds like a lot of work, and it is, but since most are already working hard on safety why not apply your time and energies to those technologies that have been proven by research to be the root foundational reason zero injury cultures have come into being.

ZeroInjuryInstitute.com can assist you with all the tools you will need to understand and apply these zero injury technologies to your business environment to create your own culture of safety where injury becomes a very rare event.

So the question before most leaders in industry today who are struggling to stop useless injury is simply this; "Are you ready yet?"

Are you ready for positive change? Are you ready to use the research proven "Zero Injury Safety Leadership Concept." If so contact the Zero Injury Institute at www.zeroinjuryinstitute.com.

- Global Airline Industry Program
  http.//web.mit.edu/airlines/analysis/analysis_airline_industry.html

Chapter 14

Safety is the Fiber Optic Thread That Ensures the Success of your Company

Safety Fiber Optics

Those contractors who are effectively using the Zero Injury Safety Leadership Concept have found many times over that record numbers of hours-worked can be achieved without an OSHA Recordable Injury. The very best achieve 1,000,000 work hours and more with zero recordables; with two contractors achieving over 2,000,000 hours without a recordable and one that achieved over 4,500,000 hours with Zero OSHA Recordable injuries.

These contractors without exception testify that their zero injury jobs are always their most profitable projects.

Those learning of this have an immediate question.

"Why might this be true?"

To explain this truth I would like to use the illustration that a project operational/organizational matrix represents a fabric; a "fabric" of skill and system utilization that ends with the completion of a successful project.

It the objective of this article to cause those who are not yet using the Zero Injury Safety Leadership Concept to look into the details and go on to embrace the concept.

Fiber-optics in wide use today in the communications industry is not a new innovation. The guiding of light by refraction, the principle that makes fiber-optics possible, was first demonstrated by Daniel Colladon and Jacques Babinet in Paris in the early 1840s. Fabric refers to any material made through weaving, knitting, crocheting, or bonding. The entomology online dictionary sources "Fabric" as "building, thing made" and evolved from "manufactured material" in 1753 to "textile" in 1791. Visualize if you would a fabric, made of woven thread, vertical and horizontally disposed, each thread crossing the others in an alternating fashion to form a textile. As one can easily see a completed project does indeed fit the description of "building, thing made" hence the word Fabric is a sound concept as I am using the word.

Visualize a Construction Project matrix, the people and controls required as illustrated below.

Systems/materials/procedures/processes/plans/tools/safety

Communication

People

Ops Manager

Proj Manager

Superintendent

General Frmn

Foreman

Craft

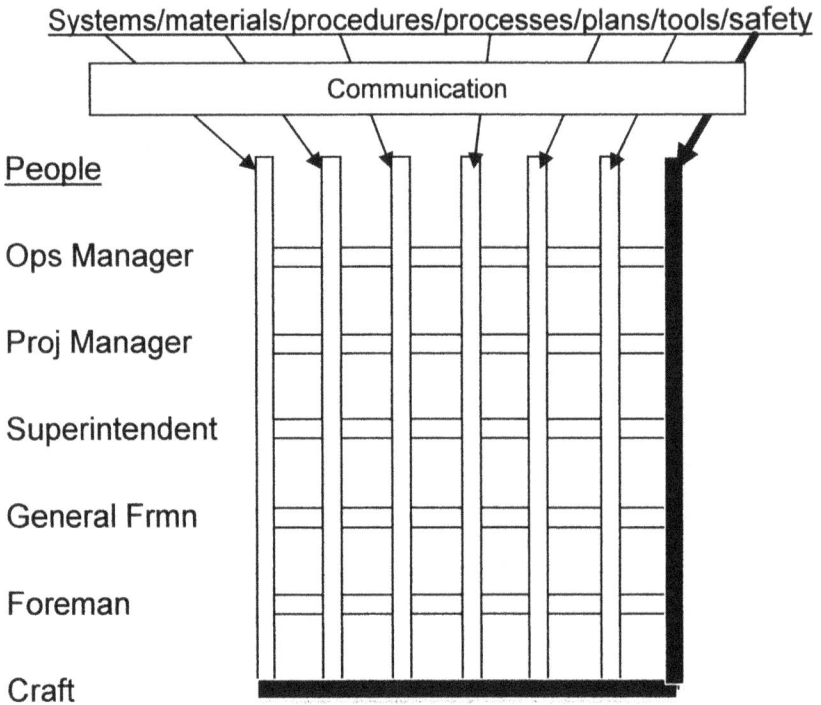

The intersections of the horizontal bars with the vertical bars indicate that all project employees have a role in ensuring all things go well with each person/position having one or more work execution roles. For each work execution role to be fulfilled there must also be a continuous flow of information to/from the horizontal bars along the vertical bars so that each individual knows exactly what their individual roles are. Each organizational position inputs their information into the matrix directed to whoever needs that information.

Two very different challenges are faced when a project is to be constructed. One is Project Engineering, the other is Human Engineering*. To accomplish a successful zero injury project it is necessary to bring these two challenges into perfect harmony.

The operational elements flowing along the "vertical bars" in the above matrix, requires a common system execution skill set one could call Project Engineering. The issues faced are typically around the "what and when" of the work. In this illustration Project Engineering is composed of Systems/Materials/Procedures/Process/Plans/Tools/Safety. These few are used to illustrate "traditional" project "implementation tools" and when fully listed typically vary little from one employer to the next.

The operational communication that must flow along the "horizontal bars" in the above matrix requires an application of a collective skill set by the listed project personnel and this could be called Human Engineering*. The issues faced are typically around the "who and how" of the work. Human Engineering recognizes that obviously people must cooperate to perform the work. This communication is a necessary "traditional" project "tool."

With the matrix in full operation the typical project will spend adequate energy to ensure the project's implementation strategy is fully engaged and that all safety considerations are in compliance with the OSHA statutes. ZII terms this safety compliance effort the "hard side of safety;" the scaffolds are built in compliance, the tools are in compliance, the electrical service is in compliance, the required personal protective equipment is supplied in compliance, etc.

ZII experience has shown that all employers attempt to define and describe the specific and individual roles each person/position must fulfill in order to ensure a project stays incompliance, on schedule and completes on budget.
But it is also ZII experience that employers have a very difficult time delivering an ideal "on schedule, on budget" project. Many unexpected circumstances arise and valuable time is used correcting the often morass of unwanted

events. All this unwanted activity takes its toll on the schedule and budget often causing the project to complete behind schedule with marginal profit and too many times the project will actually lose money.

Now back to the Question of –

"Why might this be true?"

It is because of the absence of an effective the "Fiber Optic" Clincher.

The CII research has proven the value of leaders going beyond safety compliance into the "soft side of safety" emphasizing research proven safety interventions in a number of key employee "friendly" ways. When this is done not only does the project finish on schedule and on budget with no injuries, but overall the zero injury outcome yields the most profitable projects.

One of the key features of such a project is the leaders have fully defined in detail the "user friendly" safety roles of each position so that all have a clear understanding of who is responsible for the many features of safety and thereby defines who is to be held accountable for each safety feature.

Then there is a second question people have about others achieving zero injury.

"What is it about a zero injury outcome that results in more profit?"

In interviewing the employees of those who are successful in achieving zero injury I have found at least four reasons;
    1. The craft employees state unequivocally that "This is the safest project I have ever worked on;"

138

2. The supervisors of the work where zero injury is being achieved state that all the employees are more productive;
3. The cost of injury is gone and with the injury;
4. The indirect cost of injury is also gone.

When this information about employees being more productive is shared we hear a third question.

"What do those managers do that gives those results?"

Before I answer that question please allow me to establish the traditional operational base for project safety.

Traditional Base

Prior to project start, safety behavior rules and procedures are established, published and employees are instructed and warned; comply or else you will be terminated. These rules and procedures are given to the employees in the safety orientation sessions. Rule and procedure enforcement responsibility is normally given to the Safety Specialists while the line management (Project Manager, Superintendents, General Foremen and Foremen) tend to the challenges of getting the job done; called "production." Safety personnel teach, coach, urge and finally demand compliance to all safety rules and OSHA compliance. The field supervision, Foremen and General Foremen get involved with safe rule compliance only when the Safety Specialists wish to have discipline administered for some safety infraction of a rule or procedure. This describes the "traditional;" Safety personnel in charge of safety; and line leaders in charge of production.

Zero Injury Outcomes

Contrast this traditional safety culture approach to one where the project leaders become involved in the zero injury culture building approach to safety and are held specifically accountable to perform the work injury free.

In a zero injury culture the project leaders are in charge of not only production but also safety oversight and the responsibility to ensure the use of safety planning, safe work processes and safe practices. Line supervisors oversee safety through close attention to the planning of safety, ranging from pre-project safety planning, to pre-planning safety for any significant events such as engineered crane lifts or major hardware installation and even to pre-task safety planning by foremen and crew.

In summary, in a zero injury safety culture the line leaders are responsible for safety rule and procedure enforcement with the Safety Specialists specifically forbidden to administer discipline. The Safety Specialist acts as the safety ombudsman while providing safety compliance coaching to the crafts.

Yet, even in a user friendly safety environment, in the final analysis, if an employee insists on engaging in at-risk behavior, even after careful coaching by the Safety Specialist the problem must be corrected so action is taken by the leadership after advice from the Safety Specialist. Persistent known safety non-compliance by an employee cannot co-exist with a zero injury safety culture. Employees known to engage in at-risk behavior, in effect choose not to work here!

Now back to the answer for the question of-

"What do those managers do that yields those results?"

In this case I find there are at least eight reasons -
1. Working at zero injury requires communication that is caring,
2. Caring is contagious; if you care so much for my safety then I will care an equal amount or more for your safety, and
3. Caring catalyzes a feeling of safety culture ownership,
4. Caring allows the safety culture to exist in the absence of fear.
5. Working at zero injury requires craft involvement in defining and assessing the safety culture,
6. Craft involvement creates Safety Culture ownership which creates a sincere willingness to avoid at-risk behavior,
7. It is the craft ownership of the safety culture "turns the lights on" in the fiber optic matrix concept and
8. The avoidance of at-risk behavior by all employees, leaders and workers alike, is the principle contributor to the possibility that the project can complete injury free.

Those who integrate Zero Injury Culture Building Safety into the project execution matrix suddenly find that Safety does become the light filled fiber optic thread that cements the fabric of the matrix into one homogenous blend of injury free activity and relationships.

Gone are the traditional approaches to safety where projects use a strong punishment oriented "my way or the highway" demanding posture on safety. Present in the culture is a top leader led relationship building approach reflecting an attitude of respect for others, courteous in conversation, cooperative in problem solving, conscientious in work ethic and studious in receiving and providing critical information.

When you have such a caring injury free culture in operation, if one will simply stand back and listen carefully, you can almost hear the musical harmony of a well written symphony. "Harmony" as this Symphony Score is created before your eyes as all work together to produce that product that all want in the first place, the work done well day by day, and at the end of each day all go home unharmed with a feeling of "Hey, we have a safety culture that works; and we have a project that is on time, on budget and greatest of all free of injury; **ZERO.**

- "Human Engineering" in the sense the author is using the term means the art of overseeing the effective use of people skilled in work execution using effective systems such as Systems/Materials/ Procedures/Process/Plans/Tools/Safety.
- ** The Zero Injury Institute operates with the research of the Construction Industry Institute supplemented by the 20 years of safety consulting experience of Emmitt J. Nelson who served as Chairman of the CII Zero Accidents Task Force.

CHAPTER 15

AUTHOR'S RECOMMENDATIONS

Critical Implementation Priorities

From my observations of successes and failures over 20 years of consulting, I offer the following observations and critical recommendations to those seeking Zero Injury as a working cultural norm in their companies.

1. Nothing is more important than an intense "passion" for "zero injury" at the very top – CEO/COO-management level in the company! Talk the safety talk. And "WALK" the safety talk!

On a personal basis being committed to zero injury is demonstrated when little sleep comes on nights following an injury to one of your employees. When you learn of an injury and that sleepless night arrives, then you know you are committed. And then you will conclude that the human suffering has to stop and it is up to you to stop it. It is then you begin to realize that injury to one of your employees has become "unacceptable."

But even so some leaders keep postponing the safety culture change they intuitively feel is needed, thinking that things are okay. Too often, however, I have seen the waiting end with a fatality to an employee. It is then that sleepless night does arrive in a way that was never wanted. It is too late for the deceased but not too late to prevent the next and the decision to change is made.

Think of this definition of "commitment" on the part of top leaders.

> *"If your rest pattern is not disturbed by the fact that an injury occurred to one of your employees then you may need to work on your definition of your commitment to the safety of your employees."*

2. Remember your people will take chances out of a sense of loyalty to you and the company if you do not spell out your zeal for "zero injury." The top leaders must make it clear to all that "risk-taking" in the interest of production, schedule, or cost is not acceptable.

3. Set the expectation of Zero Injuries in your company. Management will need to meet with the employees and explain why. Enlist your employee's participation. Use employee involvement processes.

4. Ensure foremen and other line management are totally responsible for safety performance. Safety professionals are advisors only. They are important. Have them working on OSHA compliance, safety training and auditing. The successful companies are holding line management accountable for safety performance.

5. Begin every meeting of the senior management and lower management groups, begin with emphasis on the safety performance of the company. Routine and frequent reports of the safety performance status are given the CEO/COO. When the CEO/COO is asked how long it has been since the last injury you always get an immediate and definitive answer!

6. Ensure that all people, at all levels, understand that safety

is of paramount importance; that it is a value, thus not subject to prioritization and certainly never should be merely co-equal with production or schedule or cost. Many of those who embrace the Zero Injury concept say that the Zero Injury Initiative actually improves these important factors. And all those who reach zero recordable injury say that the zero injury projects are the most profitable as well.

7.Expect that accident and injury reporting be immediate and have the highest profile. For instance, the CEO gets an immediate personal call from a selected management level when a lost workday injury occurs.

8.Set up an immediate job site visit by senior level executives of both contractor and owner, occurring no later than the day following a lost workday case, to review what occurred and plan steps to prevent further deterioration of safety performance.

9.Have an innovative CII research containing safety program in place that is in a continuous state of improvement at all times. Continuous improvement keeps the program before the employees and keeps their attention.

10. Consider the use of recognition awards for safe behavior performance. Ensure that it is group performance focused. Recognition for "zero" injury achievement must be given the highest level of management attention. Routine means of recognition must be developed. Monthly recognition is a good idea with annual awards for achievement also routine. The Owner or CEO passes out the annual awards!! This can take the form of a celebration dinner. Now you are walking the safety talk!

Far too often I see employers achieve zero injury and fail to recognize the accomplishment if at all and then too long after the milestone was achieved.

Recognize the achievement of zero injury and make it timely!

11. Develop means to ensure the costs of safety non-performance (injuries) are charged to each project or department before the profit or loss is calculated. Include any corporate level indirect costs. Such a system clearly tells the manager who is ultimately responsible for injury and that unit managers bottom line will carry their own injury burdens.

12. Ensure that management down through the foreman to the worker is at least annually evaluated on safety performance along with the other critical evaluation factors.

13. If subcontractors are involved, require well run safety program CII research containing content and a commitment to zero injury from them as well.

14. Utilize first day, in depth, safety and job orientation of all new hires on construction jobs. This is a good time for new hires to meet the project management personnel who can use this time to impart to the new employees a sense of urgency in maintaining a "zero" injury performance level. Insisting that Managers participate in the new employee orientation is how you effectively communicate the zero expectation to the worker.

15. Train employees in safe work habits; formally.

16. Train both foremen and superintendents in leadership supervisory skills and safety. Ensure they buy in to the zero injury expectation. Also hold these line managers of production accountable for any injury that occurs.

17. Refer to the Construction Industry Institute Zero Accident Research products. The results of the 1993 and 2001 and 2002 research are available.

18. It is very important to ensure that Line Management take on the responsibility and accountability for a safe workplace and a safe work process. In fact, the leader of a safety function in a company should be named the Safety Director rather than Safety Manager, to prevent the idea that somehow the Safety function "manages" safety. Ensure that "safety management" is first and always a line management function.

Caution urged for Monetary Safety Incentives

Monetary incentives seem to have an attraction that is deceptive. Some want to use incentives first, hoping that this money will somehow persuade employees to work injury free. This is but one of the problems opponents see in the use of the incentive technique. If you decide to use incentives do so only after you have successfully incorporated all the other Zero Injury techniques into your safety management process. In my experience it is a proven fact that safety incentives in absence of any proactive use of the proven zero injury techniques will cause employees to under-report injuries.

This is also the reason that the use of safety incentives in cash form has found disfavor in many quarters. The argument is that such incentives will, even if inadvertently on the part of the employer, lead employees to hide injury. As a result OSHA has launched an inquiry into the "hiding injury" claim.

However, some contractors remain users of the monetary safety incentive technique. All employers that do should be using incentives very selectively on projects with start and end dates. I currently know of no companies that attempt to use monetary incentives on the lagging indicator of injury frequency on a company wide basis.

If used, incentives should always be used on an inter-dependent, employee to employee format where all work to win the award or no one wins it. The object, of course, is to get all employees enlisted in the effort to cause inter-employee communications about working safely to reach a very high level.

I personally side with the contractors who simply will not use cash incentives because they are firmly against them in principal. These many that oppose monetary incentives are also being found achieving remarkable safety records. There is, of course the opposing argument that properly managed monetary incentives do a very effective job is helping employees keep their mind on safety.

PART 2

THE DETAILS OF A CONTINUOUS IMPROVEMENT PROCESS

Chapter 16

Taking Management Action

First Things First – A Detailed Approach

For Zero Injury to become a company safety culture result delegating the management of safety does not mean relegating leadership to lower levels of management. Safety leadership must be held and directed from the CEO/COO levels of management.

One of the challenges facing Top Management, however, is that of alignment on the commitment and action required to achieve a Zero Injury working culture. Such alignment can easily be achieved through a facilitated workshop where the Top Management group is guided through a process that allows alignment to occur. It would be beneficial if this workshop were preceded by an in-depth explanation of the Zero Injury concept, its origins and the logic supporting the Concept along with examples of those who are successful.

Properly managed safety, as I have pointed out, does not lower productivity; rather properly using the CII Research results in managing safety has proved to improve productivity. However it is often found that lower level managers typically do not readily accept this improved productivity concept. The reason is that it is strange to the currently in-place common traditional culture of "pushing production" that has ruled the workplace in America for the past 100 plus years.

Thus, it is incumbent on company upper management to lead the Zero Injury culture revolution to ensure change does, in fact, take place.

Once the safety commitment has been examined in-depth and redefined, the next step is to flesh out the safety program incorporating the CII Best Practices. The following is a list of the important leadership initiatives.

They are –
1. Establish a Corporate Safety Committee
    a. Led by the CEO.
       The agenda is changing the culture.
    b. Conduct a culture change workshop to develop a prioritized agenda.
    c. Conduct regular and frequent meetings on how the prioritized agenda is to be implemented.

2. Embrace Behavior Based Safety for the Manager cadre.
    a. Design a corporate accountability model.
       i. Who reports the incident to whom when an injury occurs?
       ii. Who investigates injuries?
       iii. Who is responsible for corrective action?
       iv. How soon and when are investigations conducted?
       v. Who reviews the investigation reports?

3. Install an injury management program.
    a. Develop a Physician liaison process.
       i. Create a Physician and medical service network.
    b. Develop return to work policy.
       i. Include a "Restricted Duty" capability.
    c. Assign responsibility for maintaining injured worker contacts.

d. Assign responsibility for insurance carrier interface.
   i. Audit injury reserves.
   ii. Close cases as quickly as possible, insuring the injured has recovered.

4. Develop safety teams at lower management levels as the organizational structure indicates. Ensure that these safety teams are linked. The leader of the teams just below the Corporate Safety Committee is a member of the CSC. Link lower level teams to the one above in like manner.
   a. The agenda for these safety teams is the management of safety, setting safety action goals and guiding the culture change process.
   b. Conducting zero injury workshops to create the prioritized agenda for each team is important.

5. Institute a "Top down guideline, "bottom up" goal setting process." (This process is discussed in detail in the pages following.)

Defining Safety Goals

These goals set by each safety team are not just classical goals for some number of injuries as tradition would have it. Rather they are the goals for CII Zero Injury safety technique implementation. The safety techniques found most powerful in the CII research were (See Chapter 9 and 10 for details):
- Demonstrated management commitment
- Staffing for safety
- Safety planning
  o Pre-Project planning
  o Pre-Task planning
- Safety training and education
  o Safety Orientation
  o Formal Safety training
- Worker participation and involvement
  o Safety teams
  o Behavior Based Safety training

- **Recognition and rewards**
  - Weekly
- **Subcontractor management**
- **Accident/incident reporting and investigations**
- **Drug and alcohol testing**

Chapter 17

The Process That Yields Evergreen Safety Progress

An Integrated Safety Management System

Now that you have the background information and safety techniques and the detailed approach to be used to achieve success, let's review an employee-involvement Safety Team process.

As stated earlier, all the work in the world will not yield the result of Zero Injury unless you have a process that involves the employees.

Not providing means of involving the employees/workers is the traditional approach to safety management.

Those achieving success today in eliminating injury are using employee involvement processes. I have discussed Behavior Based Safety above. Now I will cover in more detail "safety teams," for that is where my experience has been.

The Safety Team is a process that, if followed, will lead you and your organization to that coveted goal of "Zero Injury." It is a pathway that has been blazed by others. The landmarks are clear. Such a path is revealed in the following pages.

While one will find a few references to specific works by others, the information for this book, in addition to the author's personal experience, has been countless sources on the subject of managing safety. These data sources have been integrated throughout four decades of practical application, and in the last few years, took on the present form of what is termed in the book as an "integrated safety management system".

Few companies bother to use an "integrated system" approach to safety management. Most seem to use a series of tools, or sub-systems, that manage safety plans, techniques, costs, production, and completion schedules. All these are needed, but there is a failure to integrate the skills of the workforce, in an optimum systematic fashion, to ensure a mode of "automatic safety progress" that is pervasive and ongoing.

Automatic Safety Progress

Those that install the system explained herein can acquire that illusive competitive edge of "excellence" that typifies those that are the best in their business: those whose employees are rarely injured.

The "safety management system" presented to you in the following pages is not a theory on how to manage safety. Rather it is a proven system, tried and tested prior to being recorded here. It is not a quick fix; rather, it is a permanent fix.

The system installation process takes time and hard work. However, once you have the system in place, your "safety management search" will be over. You will be well down the path of acquiring the feeling of achievement that goes with a Zero Injury working culture.

This safety management process does incorporate many of the management techniques and tools espoused by various safety management authors during the last twenty years. Those included are techniques and tools that work, not only in principle, but in fact.

The approach used in this book recognizes that most skilled managers do not have time to read a mountain of information to get the essence of a subject, nor do they want an excess of help. They are anxious to use their own talents and innovation in filling in the fine details of how to design and implement a system.

Thus, the object of this book is to describe the broad basics of this proven safety management process in almost an outline form. You will find the system elements described in brief fashion with the attendant sacrifice of a lot of fine detail.

I do not try to overwhelm you with persuasive arguments. A little logic is presented and then the system element. Herein are the bare elements of a proven effective safety management process that produces the elimination of injury "automatically."

"Automatic" does not mean that no effort or energy is required. To the contrary, a lot of energy will be required to install and implement the system. The important thing about this energy, however, is that it brings about progress. And in today's competitive business world, one must make significant progress on a continual basis in order to remain in a competitive position in the market place.

People Involvement

Involving your people in the creative safety management process is where "the system" derives its strength. Safety Teams are formed to address safety issues and solving the question "How do we become a Zero Injury culture." These teams involve more than just your managerial and supervisory cadre; they also involve all your people, including any on an hourly payroll.

Embarking on a "safety team people involvement" process is best if done as a part of a long-term commitment. Long term is because, once experienced by your people, the mass of them will become so devoted to the process that leaders can demoralize their organization if they to revert to a totally authoritarian style.

The fact is, you will not want to revert for, you will see the power and the resulting world-class safety performance. If you take the commitment to utilize safety teams, do so with the knowledge that, in many respects, it is a "one-way path". It is, however, a "one way path" to Zero Injury. A worthy end!

Having experienced the feeling of achievement that results from being a part of the installation of an "integrated safety management process," and having shared the resulting rewards, it seems mandatory to the author that the system be recorded for others to use.

Good luck! It is hard work, but the results will be worth it.

Choosing A Name for the Organization (for this book)

Enterprises that can benefit from the installation of a safety management process come in many shapes and sizes. In general, the "names" of these "work places" or "work efforts" come in an equal number of choices.

For instance, the following are a few of the names that apply to these places or enterprises: shop, mill, company, office, firm, corporation, department, division, region, project, staff, mine, facility, factory, store, refinery, bureau and plant.

"Facility" was chosen to identify the "work place" referenced in this book because it was in a manufacturing facility that the system was proven. The point in selecting a name is unimportant other than to point out that where people gather in an organized fashion to pursue a common objective in our "working" world varies and does not matter in the final analysis.

The presentation of the conceptual elements of the "integrated safety management system" is done in Chapter 16 using just ten steps and is free of extensive, in-depth explanations. This does not mean that the various distinct aspects of the safety management process presented fail to embody a degree of complexity.

During the author's experience in installing the system, a multitude of questions were generated by those involved that required answers. As these questions surfaced, they were typically answered via articles written for wide in-facility distribution. Rather than include all that detail in the body of the book, those articles have been appended. Each deals with a variety of questions that go beyond the content implied by the title.

CHAPTER 18

SAFETY MANAGEMENT SYSTEM EVOLUTION

The Beginning

Since the beginning of the Industrial Revolution, managers have been learning how to manage safety. One observation seems to reflect the nature of the evolution of safety management expertise.

"That which worked yesterday somehow is not quite adequate for today."

Safety management professionals have continually worked on the latest applications of the most advanced theories. Volumes, and more volumes, keep pouring out of printing presses. Recently, the subject has been Behavior Based Safety. Most all is extremely good material, along with concepts with which to work, but none give the answer sought on how to achieve "evergreen" progress.

There also seems to be a recent correlation that reflects that the more brief, succinctly ("One Minute" & "Idiot's Guide") written materials win large enthusiastic management followings. I have not seen a publication named "The Dummies Guide to Safety Management." But wait awhile, it seems there is one for any conceivable challenge. In time, someone will publish something on safety with such a title.

What seems to be lacking in publications available today on safety management is a process that integrates all the material available. Lacking is an overall plan that assembles this mass of method into a "system" one can use to make progress; an "integrated process for safety management" is needed that allows the broader spectrum of the management challenge to be effectively addressed.

In dealing with the challenges we managers frequently face, we run to our safety management toolbox and find that we really do not need a specific tool, but a way to simultaneously use many tools.

It seems to me we are in need of "a safety management process."; a safety management process that, once present in the organization, will yield progress in reducing "at-risk" behavior on a continuous (automatic and ever-green) basis.

It is not merely the same progress that we now strive for in an authoritative fashion. It is an even higher level of progress that a well-integrated safety management system, using people involvement, will automatically give you.

In fact, this "progress" to which I refer can become so automatic that, once you have the system working, you will sometimes find yourself watching in amazement.

CHAPTER 19

SAFETY MANAGEMENT CHOICES

Zero Injury Team Choices

In the technological and industrial world we live in, many options exist on how to manage safety in an enterprise. To achieve a Zero Injury working culture, the really important team choices are made early.

The team decisions you will need to make are influenced by two basic situations.

One, you have a short-term management challenge, such as a project where you need to involve people at all levels in creating the Zero Injury safety culture.

Two, you have a more long-term situation where there will be low employee turnover. In this later case there are again two sets of circumstances.

One, this is a new facility; these are new people and the opportunities on how to structure the "people" part of managing safety are almost endless.

Two, this is an existing facility; these are in-place people and the opportunities are, at best, limited on how to devise a safety management system with which to work.

If your situation is the first, you can enlarge your staff by hiring a safety professional or, if you choose the more temporary route, you can retain a safety consultant and carefully design your beginnings in safety management with teams. You are indeed blessed. If you heed sound advice, your success with safety team management can almost be assured.

This implementation section has much information for you within its pages, and you can use it while you do the basics on how you want to involve your employees in achieving the Zero Injury culture you desire. A few choices are Teams, Quality Circles, and Self-directed work groups. What will it be? You'll still need a sound safety management process, but you can do the structural basics simultaneously.

If your situation is the second described above, "this is an existing facility, these are in-place people and my structural opportunities are limited---" this book also has much for you.

Looking at this second type of work place, on the surface, your people seem to do the tasks they are assigned fairly safely. These tasks have become "their" jobs. You sense they do their jobs well and they want to be left alone. But you find injuries occurring. You wonder why?

As boss you agree. "Let me leave them alone," you say as you think this through. "They will be happier, they will get the job done and I'll be happy," but if only we can achieve a shift to the better in safety performance.

Lets' say you do wait, thinking it through, looking for options. In fact, you do it well. But after a while you notice your people simply aren't making progress in safety; at least, not enough to meet the competition in the market place, and satisfy your own high standards to be the best.

At this point you may be tempted to intervene. "I'll ask for more safety focus," you think. But deep inside you know there isn't much more focus to be given. You fear if you push much more for more, you'll get less: because, due to the pressure from you, your people may start to do unwise things that will result in increased injury; not less. You want them to think of safety as part of production. You want them to realize that improved safety will not sacrifice production, but will actually improve production to be safe production! It is likely that if you told them this last fact, they would not believe you. Somehow, you want to create a process that will, in it-self, prove your position without their being told. A way needed to improve safety so, when it is over, they will realize that productivity, was not sacrificed but rather was improved.

Faced with this challenge and your experience, you have been to the safety toolbox many times in the past, and tried many programs. It seemed that some helped, but soon the place was back to needing another safety technique; another shot in the arm.

How then can one make this progress toward Zero Injury, in a situation like this?

The answer is simple: Get help.

You ask, "Get Help? I thought we were leaving the tradition behind," you answer. Help from whom?"

You have asked the key question: from whom?

If I told you to ask for help from all those from whom you are reluctant to ask for more; those that are doing their job and doing it well, but leaving safety behind too many times; what would you say? - or ask?

Perhaps your question is "How can I do this?" If you

sincerely have this question in your mind, then, you are ready to start down the path to safety progress to an even higher safety performance level termed "automatic safety progress" that will guarantee your future as a Zero Injury culture.

Chapter 20

The Safety Management System

The System

The Safety Management System consists of ten steps that if followed faithfully will accelerate your chances for rapid progress toward the establishment of a Zero Injury culture where worker injury is a rare event.

These ten steps are:
1. Setting the Safety Foundation
2. Getting Help for your Safety Culture Change
3. Using the Safety Team Help Process
4. Creating Safety Action Goals
5. Using System Operational Guidelines
6. Using the Safety Goals Review Process
7. Capturing the Progress as Evergreen
8. Working Out Relationships
9. Making Quality Part of the Process
10. Invoking Progress Now!

STEP ONE –

SETTING THE SAFETY FOUNDATION

The Most Important Ingredient

Most managers today agree that the most important ingredient of an enterprise is not the hardware, nor the money but the people. One could call the facilities and equipment the foundation.

This simply recognizes that the very latest in facility equipment with plenty of money will not automatically yield safety success. These are essential but the paramount "key" to the success of your enterprise in safety, to reaching the coveted Zero Injury performance and yielding automatic progress is in how well your people function in recognizing and avoiding "at-risk" behavior.

This being the case, then you will need to spend significant time examining and rethinking how your employees are treated as a part of a revised Zero Injury safety management effort.

Start With the Basics

We'll start with the basics, the safety management foundation, if you will. One cannot build a lasting structure upon an ill designed or inadequate foundation. Where or how then can we find the material for this foundation? Since we are going to consider basics in creating this safety management system we will need to consider the fundamental purpose of the facility we have.

Let's call this "fundamental purpose", the facility mission.

What is the mission; what commodity, service, product or products is this plant supposed to produce; and with what broad perspectives in mind? In addition consideration is required in how you wish your people to function including their health and welfare.

It is recognized that, in some cases, the "mission" definition has been written by some other higher authority. Your job as leader of a sub-group is to guide your people in adopting the higher mission statement and operating within its boundaries. The following information is for those who are in a "mission creation mode."

Have you made it very clear in any mission statement you have in place that employee safety is of paramount importance. Is it clear to the employees that you are not the least interested in making a profit at the expense of their well being? Is it clearly stated in your mission statement that an injury to any employee is simply not an acceptable event in the course of performing their work?

"Oh, that's obvious," you say, "everyone knows that." Well, if everyone knows, why then can we not continue to make the safety progress that we need to remain competitive? Why do employees insist on taking shortcuts and utilizing "at-risk" behavior in order to accomplish their tasks? Might these "at risk" behaviors have management's tacit approval?

So bear with me. Think a restated mission that includes employee safety as a major aspect of a successful corporate cultural change effort. Think of it in terms that use 100 words or less. Think of it in broad terms, not in great detail. We'll get to the detail later.

Mission Statement Includes Safety

Create the Mission statement draft and include appropriate references to safety so that the value tenets covered become the underpinning of your organization. You might even want to, indeed you should, involve your immediate management group to help you create this statement. Use the group that reports directly to you. After all, they need to experience this thought process also. Together you address the question "What is our corporate commitment to safety? Are we only "really committed?" Or are we going to be committed to the point that "it is not acceptable for an injury to occur."

Remember that this Mission statement you are about to create is the foundation on which you are going to build this safety management system. It must be a statement that you believe and support and are willing to publish for all to have. And just as importantly, you are prepared to "live by the mission words." What belongs in The Mission Statement? Use simple statements. Include the basics.

For instance:

"The XYZ Company's purpose is to produce the best 'Zaphs' on the market. Zaphs manufactured here will appeal to our customers and fulfill their expectations.

In order to accomplish this we wish to create a working culture where it is clear that our employees are our most valuable asset, and our customers our most important concern. Thus it will be important in our mission to protect this employee asset against injury or abuse. Since employee injury is not an acceptable event in XYZ Company we desire and will strive to create a working culture where there are zero injuries to our employees. In so doing we feel we can more properly serve our customers.

We also wish to conserve our facility asset. All should recognize that in order to secure our stockholder's investment in the facilities, and our employees continued employment that a profit is the anticipated result of our efforts."

Be honest. In a free enterprise system, motives are known anyway. Spell out the important relationships that need to exist. It is the information contained in the Mission statement that will become the common safety objective that will bind you and your people together as you travel this newfound pathway to an injury free culture.

Set Your Working Climate

A safe working climate (safe working environment) is of paramount importance. When your people know what their collective "mission" is, they also need to know something of your broad expectations regarding the "safe working environment." Most leaders, by not sharing with the employees their visions regarding safety leave this important "workplace condition" to chance.

The benefit of thinking through the elements of a "safe working environment" tends to solidify how one acts around the workplace. Developing a "Safe Working Environment" as a part of the overall Mission document commits the leaders to certain safe behavioral and management norms that they individually and as a group must consciously support.

Some have called this an "organizational climate for safety." This desired safe working environment, or climate, needs to be defined and written also, again for all to see. This should be no more than a one-page document that describes the "people considerations" that you as manager would like to see as work place norms.

For instance:
- Employees will have opportunity to gain personal safety information to help them in avoiding "at-risk" behavior.
- Employees will reflect a friendly and cooperative attitude toward their peers as they work on a safe workplace together.
- Safety information needed to perform the tasks to be done will be available when required.
- An attitude of "pride in safety" will be prevalent in the facility.

Most of the lists I have seen contain 12 to 15 statements of similar nature.

You can look at the Safe Working Climate document as a loose framework of safety norms that you would like to see in your company or operational segment. If at all possible it is important to allow your employees to be a part of the creation of such a list. If they are involved in the creation process, the safety norms will be largely in place when the document is finished. Otherwise, it will take time for the work force to adopt those created by others as their own.

STEP TWO –

GETTING HELP FOR YOUR SAFETY CULTURE CHANGE

Making Progress

Progress never comes without change. You can print it, frame it and hang it on the wall. It's true, has always been true and always will be true. Accept the fact.

Progress never comes without effort. You can work doubly hard and drag more "effort" out of everyone or, better still, you can get this extra effort to voluntarily flow.

Progress never comes without involvement. If you mandate progress, the people affected are involved. If you catalyze voluntary effort, you also get voluntary involvement. And I ask you this, "Which is more powerful; mandated involvement of voluntary involvement?"

Think "change, effort and involvement."

Then think of "how." How do I instill in the people on whom I now depend an "attitude" that will yield automatic safety progress?

You do not do it by telling them that they need a new safety attitude. It is not a "telling" they need.

You start by "asking". But, it is very, very important that you ask in the right way and that your reasons for asking have a sound logical base.

The Nature of People

In order to develop this sound logical base, let's deal with the nature of people. Research has shown that it is the nature of people to want to be involved to some degree in the decisions that are made that affect their work, work place safety, and the work product.

To begin with then, we must realize that to "ask" is to "involve." When people become involved they expect some degree of response to their answers to the asking.

So you must use a process to involve people that "asks" them in the right way and that will deal with the responses they give in a responsible manner that gives them feedback.

What kind of process? What are my options, you ask?

STEP THREE –

USING THE "SAFETY TEAM" HELP PROCESS

Automatic Evergreen Progress

Remember, you are working toward a safety management process that yields "automatic evergreen progress." So when you take a step toward installing a part of this safety management system, remember that it is not a "one-shot" effort. What you do is not only for now but for the future as well.

You are going to develop a system that is integrated, in place and yielding progress. You also want progress despite the periodic perturbations that face the organization, a new government safety regulation to meet, a new product or product safety specification to meet, or a change in safety staff from time to time.

So the Help Process that you create should be one that continues, evergreen in nature, and which is flexible enough to manage necessary change. The current safety management toolbox contains a number of "involvement" options; Quality Circles, Task Teams, Safety Teams, and Behavior Based Safety, to name four.

Quality Circles have been used to involve employees in a quality of product sense, in a quality of work life sense and a safety sense.

Task Teams usually have a finite life and address a specific "short term" issue.

Safety Teams

Safety Teams are a very powerful involvement process that shows strong promise for the long term. Teaming safety is a powerful process that allows employees to set their own safety goals, from the bottom (crafts) up. (Not injury goals, but goals in how to work with less "at-risk" behavior.)
When one considers teams please realize teams can have considerable variation in sophistication.

In safety one can begin with a very elementary team. Elementary teams are recommended for a workforce with high turnover. Sophisticated teams require more meeting time and investment of training thus are recommended for a more permanent workforce.

STEP FOUR –

CREATING SAFETY ACTION GOALS

How To Create Goals

Knowing "how to use" a Safety Goals Program is just as important as creating the program. A Safety Goals Program will often yield little, if any, progress if the program isn't used in the right way.

It is productive to set up an understanding on the meaning of and use of words. My suggestion is that "Mission" be the overriding and guiding document that allows Objectives to be set that are aimed at achieving the mission. Specific goals are set to accomplish an Objective.

Let's talk about the "nature of people" again. When it comes to people and goals there are two kinds of situations.

In the first situation, the leaders develop goals and pass them down to the employees. These are top-down goals and. while usually accepted as "how 'they' do it around here," such goals very frequently have only half-hearted support from those effected and the level of commitment, which might seem satisfactory from afar, when viewed up close is found to be vague and uncertain. Questions from the employees sound like "How could they arrive at such a goal; don't they know what's going on down here?"

The above underlined words are the key explanations of the problem. They, they, and down. Passed "down" from "up there."

In the second situation we find "the way" to implement a safety goals program. This "bottom-up" method will often yield more stringent safety goals with attendant higher performance results than could be exacted with a "tops down" approach.

Guidelines Only From Management

Start by thinking "Tops Down Guidelines" - not goals: only guidelines. These guidelines will be the "direction setters" for your progress. In each area that you desire progress include in the Safety Goal Guidelines statements to that effect, i.e. Guideline:

1. Each safety team is to develop its own safety mission statement toward the achievement of a Zero Injury work team.

2. Each safety team will set stretch safety goals that call for implementation and adherence to safe work techniques.

3. Each safety team will set stretch safety goals that will reduce the amount of "at-risk" behavior found in the execution of work procedures and processes.

4. Each safety team will develop goals that will result in improved knowledge in safe working skills and team skills.

5. Each safety team will forecast, for budgeting purposes, the expected training required to accomplish its safety goals, including any capital expense.

You might ask, "But what about costs? Won't they skyrocket? How can I, as leader, on the one hand give people the responsibility to set safety goals, some of which obviously cost money and at the same time contain the potential cost explosion?"

Do this too with guidelines. Note in your cover letter what the business climate is, how profits have looked and ask for help. There are many ways to increase profits while not sacrificing safety. Often your people may spend more in one area and reduce costs even more in another area. In your guidelines letter inform those setting goals, that management at a level higher than the team leader will review the team goals at the outset of the new fiscal year and then quarterly thereafter.

STEP FIVE –

USING THE SYSTEM OPERATIONAL GUIDELINES

The Key to Automatic Progress

The key to putting your "automatic progress" system on automatic is in the holding of "Safety Goals Reviews." Reviews give you the opportunity to influence safety goals in their early state: to keep up on safety technique implementation progress, new developments and new safety issues. Reviews put you in close contact with the safety action on a regular basis and through careful questioning that you can gently guide.

Developing Guidelines

Allow me to structure the Guidelines, Goals, and Review process in a chronological manner. Do not be deceived. The entire Guidelines, Goals, and Review processes require a lot of time and energy. But it is time and energy that is specifically dealing with that which you desire most: a Zero Injury Company, and such a company is tantamount to "safety excellence."

Spending time on safety goals is time well spent. To not spend this time, means you are choosing to remain in a stifled, no-progress, mode dealing with the problems of today in a reactionary manner as you have in the past.

What you must do to convert from "reactionary" safety management to "pro-active" safety management is to adopt a safety management system embodying a goals and review process.

When the people in your organization set their goals, they

will necessarily address dealing with all the problems they face everyday. As these problems are dealt with, solutions are found and problems that have been historic go away. The time thus available can then be spent on preventing or setting up a structure to deal with the problems of tomorrow. You are, at this point, moving from a "reactive" to a "pro-active" safety management mode.

The chronology of your approach is very important. Start by issuing "top down guidelines." Select the time of year you wish all goals to be created, re-examined, updated, dropped or added. A few months prior to this, publish your guidelines. Include in your annual guidelines cover letter any philosophies or developments in the business that you wish to pass along to your safety goal developers.

When operating in a Safety Team mode your sub-leaders develop their departmental or functional goals in concert with their involved people. This process is begun by everyone developing ideas for their own individual work-related team safety goals, then bringing them to the larger work group "safety goal setting" meeting. At this sub-leader meeting resolution is reached on the group goals.

The sub-leader safety goals become the material from which the next level up, team's safety goals are formulated. It will sometimes be necessary to add the broader group multi-effort goals to the list. For instance foremen will bring their crew team goals to the meeting and those goals may include suggestions for a multi-foreman effort on safety in some critical arena. This then is addressed at the next organizational level.

Guideline Content

Your annual guidelines should include all the broad areas out of the Mission Statement for which you wish Objectives

and Goals to be established. A suggested breakdown might include some of the following:
- Safety Objective
- New Safety Goals
- Ongoing Safety Goals
- Safety Planning
- Safety Training
- Safety Compliance issues
- Safety Team Effectiveness

Further breakdown with specific guideline statements can be in order under each of the above topics. For instance, under Safety Goals a guideline statement might be:
-Each division or department is to conduct a departmental wide safety training seminar each quarter.

Notice that you do not tell them exactly when or what the subject matter need be but you are establishing through the guideline that such a safety seminar is an expected norm. In the goal development process the specific action steps, timing and subject matter for the seminar are determined.

The goal might appear as follows:
Goal:
Conduct a safety seminar on eye safety with all departmental personnel.

Action:	Timing Mo/Yr	Action by: Indiv.
- Contact "Eye Safety Inc." for program materials	09/2003	Joe
- Develop presentation	10/2003	Jane
- Arrange for facility	11/2003	Jack
- Present seminar	12/2003	James

Action steps with timing commitment and responsibilities are extremely important. First, these create logical steps that

180

will yield progress and two, estimates the timing of the progress. To omit action steps and assignments is to slow down progress, perhaps to "no progress."

STEP SIX –

USING THE SAFETY GOALS REVIEW PROCESS

Time Schedule (April-March Fiscal Year)

An overall time schedule based on a calendar year might look like this:

Guidelines -	Late November
Goal Setting -	December/March

Reviews:
Previous year progress
Current year goals -   April

Second Quarter Review - July

Third Quarter Review
With first six month's results- October

Develop next year Goal
Setting Guidelines-   November

Fourth Quarter Review - January

The frequency of the reviews

The quarterly reviews are of paramount importance. If your people, who have set their own goals to your guidelines, are to really believe you are interested in what they are doing, their successes and failures, then "you must review quarterly."

To review less often (reflecting lack of interest), you will find less progress. To review more often (too little passage of time), you will find little progress. A so quarterly review works out about right for most situations.

On Automatic

So now you have it on "automatic." All you have to do is ensure the Guideline letter is issued and the dates for reviews are established. It is not hard to manage after you get the process going and the momentum is in place.

You and your people will improve with each passing review cycle. The quarterly goal refinements will be added as needed by the teams and "automatic ever-green safety progress" will become second nature.

STEP SEVEN –

CAPTURING THE PROGRESS AS EVERGREEN

Doing it Over and Over

Managers know that the process of managing safety involves doing some things many times over. Safety Procedures to handle many activities, especially those that are complex in nature, need to be developed and systematically documented. These become "Codes of Safe Practice" or "Safety Manuals."

Employee turnover, if nothing else, dictates the productive nature of "documenting the how" of repetitive safety activities. The need to document is especially true where experience has taught that things seem to drop between the chairs and the "I thought you had the action" statements become the response of the day. Most organizations are fairly adept at reducing repetitive items to a procedure. Usually this is done after some recurring difficulty has proved the wisdom of such a step.

From Reactive to Pro-active

The pro-active minded organization, that is achieving at the Zero Injury level of performance, will have time to develop a crisis resolution procedure to handle a disruptive event before it occurs. This is especially important in critical safety areas of focus, but is also important in preparing for environmental issues where prevention gets very high priority.

There is a lesson to be learned here that applies throughout an organization. It is productive to develop preventive procedures on many other functional activities. The commonly used procedure of Inventory Control is an illustration of this.

It is also important to ensure that safety procedures, once developed, are given to those who need them, that they are explained (training conducted) and that they are used. Also important is to install a record keeping and filing process that allows safety procedures to be found when they are needed. Such a system should also provide a mechanism for reviewing and updating all safety procedures on a set time interval. Out-of-date safety procedures fall into disuse so it is mandatory to have a review and update process. Sponsors can be assigned each procedure to see that a timely review and re-issue is conducted.

It is also very important that the procedure manuals be assigned specific individuals so when update time arrives a mailing list will get the revised procedure to all those holding copies of the manuals.

STEP EIGHT –

WORKING OUT RELATIONSHIPS

Getting Along Productively

This Step of the management process addresses all relationships and conflicts, not only those that arise out of safety management. Most relationship problems in the work place stem from conflict arising out of "role" understandings or misunderstandings, so these need addressing.

As indicated earlier, it is very, very important to recognize when and where this conflict exists, and to actually put in writing a coordinated role element description for people who must work together to accomplish a common task. The benefits are obvious. Once role elements are reasonably well understood most, work place conflict ceases and that conflict which remains stems from such things as differences in ideology, personality and style.

Human relations research has been done in the field of helping people understand other people. Some theorists feel that the only way to work out understandings between people is through facilitated confrontation. And this is probably true if one considers confrontation as the event where the two (or more) of the people involved get together in order to promote an atmosphere of understanding. Only where rational working relationships are desired, can the people involved create one.

Do Not Try to be a Doctor of Psychology

"Fine tuning" severe relationship problems should not be undertaken by the average "lay" (in the sense of human relation expertise) facilitator. Rather, the troubled parties should seek out specialists with the appropriate training and academic credentials. These "experts" are available in sufficient number and background that one can be located with specific expertise that would fit most any need.

The question should be answered as to whether or not to take this conflict resolution step of "working out relationships." Occasionally, people do learn to work very well together. They see the need to overcome personal differences in style for the good of the enterprise. Developing a conflict solution process, one that fine tunes relationships is a productivity item. If you need to do it, and have not, how can you ask others to become more productive? A good example is the best teacher of all.

STEP NINE –

MAKING QUALITY PART OF THE PROCESS

Safety is Quality

Right away, it is appropriate to state "Safety Is Quality!" Any injury is a non-quality event that occurred in the execution of the work. Non-quality costs money. Injuries cost money; more than most executives dare to realize. Of course anyone can see the medical attention given an injured employee costs money.

As stated earlier, even more costly are all the indirect costs that accrue after an injury occurs; Lost time, lost productivity, poor publicity, are but three. Research accomplished by The Construction Industry Institute found that indirect costs run twice to twenty times the direct costs. In such a scenario an injury that has a medical/indemnity cost of $40,000 would have an indirect cost no less than $80,000, ranging to $800,000 when a lawsuit results. So having an injury free work culture is a profit protecting initiative. Therefore, employee safety is a quality issue of the first order.

The quality revolution may well be the single most significant development in American industry since the creation of the assembly line. Quality. What is it? Where can I find it? What can I do to get it? Where do I start? How do I measure it? How do I know if I have it? How do I keep it?

All are questions of the day. And so it should be. For too many decades American industry has failed to fully understand the defeating nature of a work place psychology of production first; quality, well yes, but not first. In the 1970's it seems that the entire production thrust of the United States that reached gigantic proportions became lost in the rough seas of consumer malcontent. Consumers found the product they sought was manufactured in another country, looked better, fit together better and worked better. The result was inevitable. In the 1990's American producers fought back. Juran, Deming and Crosby are now almost "household names." Where then does this need to embrace quality fit into our safety management system?

Embrace Standards of Quality

If you are following the "pathway to Zero Injury" then you must embrace quality concepts in your safety operations. First let me ask this question using Crosby's "right the first time" logic.

Have you ever engaged happily or contentedly in turning out a product that you knew full well would have to be redone when finished because you accepted that it would take you two or three times to get it right? Another example, more personal; you go for surgery; ready for three efforts before reaching success? Or even two? NO! NO! Get a surgeon that will do it right the first time.

All of us, throughout our working lives have a single, common desire by nature of our implied job charter. "Let me do this. And do it right the first time." Some students of "quality concepts" who work in research or exploratory fields where trial and error may well be accepted have difficulty accepting "do it right the first time" as a valid concept. In such cases I believe those working in these activities need to focus on the experimental process asking such questions as, "Am I performing this effort in the right way every time?" Newly discovered inventions often see several stages of development before they are ready for service.

I think of "safety quality" in the following manner. Quality, in a sense, is the "fiber-optic" thread along which we design our path of safety progress in the tightly woven fabric of our integrated safety management system. Quality is the essence of our everyday safety effort. All safety programs, goals programs, review processes, people involvement effort, role definition, and conflict resolution; are aimed along this fiber-optic path. The "path of do it once so the problem goes away forever!"

Yes, we need to accept "quality concepts" into our safety management system. We must make it a part of the safety guidelines, goals, roles, procedures and reviews. We must build it in. After all, "a quality safety product" is the sought after result and will become the strength of our enterprise.

STEP TEN –

INVOKING PROGRESS NOW!

Progress Begins When You Begin

The author is the first to admit the safety management strategy advocated in the "integrated safety management system" approach is a significant departure from past practice by many line managers. As such, the approach will breed doubt in the minds of some who read this work. But if you fall into the category of the "tool box" safety manager and have tried them all, then what can you lose?

The biggest mental deterrent you may face in deciding to try this system will be "but I need progress now! Not later."

Let me assure you, safety results will begin just as soon as you begin. The "word" will fly through your organization. "Employee involvement" has an immediate productive "ring" to it. However, if you fail to follow through, your organization will suffer more than with any other of the "efforts" you have implemented in the past. The disappointment factor in being promised "employee involvement," only to see the boss fail to fulfill his word, will produce a significant negative morale result.

Safety progress will begin as soon as your people begin the goal setting process. Progress will begin when they begin to "think" about their goals. As the goals form in their minds, progress begins. Once it begins, and you follow through with the entire system, you will be on your way.

In recent years, the author has encountered managers responsible for safety, who at a given point in time, were experiencing rapid progress in reducing injury. On inquiry, it was found that sure enough, a tool was being used that highlighted "employee involvement" with parallel support from the top to commit resources to solve problems. Almost without exception, however, the managers so blessed were very concerned about what to do next. They had not yet realized that there within their grasp was the most powerful safety management process available. All they had to do was push the automatic button by taking the steps to implement the employee involved system of "tops down guidelines, bottoms up goals" on an annual basis with quarterly reviews.

Bottom-up Goals

Immediate progress is inevitable if you adopt a "bottom up" goals activity in the lower reaches of your organization. Using the Philip Crosby "quality is free" concept to fuel your "bottom up" goals thrust will ensure the goals process will have problems to solve that yield an immediate safety return.

Releasing accountability, or control of your safety operations, is not advocated. The leader must always set the performance expectation, stay close to the action and "guide" it.

The typical authoritative managers do not "guide" his/her organizations; they "command" them. Do this, do that, all from the top. This "command psychology" will occur even if the manager does not intend to appear this way. Our American culture, to a large degree, dictates how people see the boss. And our cultural heritage has been authority and command!

"Guiding" is a much more delicate approach. Publish the "guidelines" which will set the tone, deal openly with the problems faced, ask for help and give your people the "format" for the "bottom up" help process.

A typical scenario, in the competitive business world of today, is that the top manager instinctively senses he needs help from the bottom. He asks for it only to see significant moral support from the organization produce little, if anything, of lasting consequence. Why?

The answer is because there is no system in place to allow the moral support to take on substantive form. The installation of the "integrated safety management system" is mandatory if you want safety progress on an evergreen basis.

Use the Ten Steps

Repeated: the ten Steps to form the "evergreen progress safety management system."
1. Setting the Safety Foundation
2. Getting Help for your Safety Culture Change
3. Using the Safety Team Help Process
4. Creating Safety Action Goals
5. Using System Operational Guidelines
6. Using the Safety Goals Review Process
7. Capturing the Progress
8. Working Out Relationships
9. Making Quality Part of the Process
10. Invoking Progress Now!

Customize it for your organization and you too will be amazed at the progress your people make in a very short time.

Chapter 21

Safety Teams

Elementary Safety Teams

Each employer will need to examine the organization to determine just how the Safety Teams will need to be structured to ensure the people on each team are inter-dependent. For instance, in, the construction business, where there is typically high employee turnover, the Safety Team of choice can be the elementary type. "Elementary" simply means that the training of members is limited to being informed that they are part of an inter-connected safety team goal setting process. Of course, the foreman (leader) of each team will need instruction on how to function as a team leader. For the members, the role of their team and their individual role as team members is explained.

Elementary Safety Teams for projects or construction divisions are interconnected through the supervisory staff. A "bottom-up" goals scenario begins with the first level of supervision and progresses up the organization with each leader of a lower team being a member of the next level team. The foreman and the crew are the first team. The second level supervisor (Superintendent?) and the foremen reporting to him/her are the second level team, etc.

In basic fashion, the process begins with each team leader receiving a copy of the safety goals guidelines letter. The foremen start the "bottom-up" process by involving their crews in setting their safety technique implementation goals and activities. The subject areas for a foreman and crew are limited to those areas of safety where they are in ultimate control of the activities required to implement the goals the team sets. All other activities flow along the customary lines of authority for planning and execution.

Creating More Sophisticated Long Term Teams

There are many sources of help on how to set up a long-term sophisticated team concept. It is not a simple subject, nor can it be done quickly. To the contrary, much thought must be given to the how, who and what questions.

The how: How do I do it? Here I strongly urge you get an education in the use of the "team" management tool and how it can be applied to safety. Such will help answer the next two questions.

The who: Who should be on the safety team or teams?

The what: What do the safety teams work on? What is appropriate subject matter for their agenda? On what frequency should their meetings held?

Teams are a very powerful "Help Process." Safety Teams are organizational structure tools. These teams also need tools to use, tools with which they can address the appropriate subject matter, in the appropriate manner. They will need to learn how to manage their team meetings to make them effective.

Safety Team Authority

It is a common question when Safety Teams are established

that the team members are unsure exactly what their scope of authority includes. This is an important subject and requires certain definition in order to reduce the non-productive time teams can spend discussing changes in areas outside their own scope of authority. See appendix on "What a team is and Is not."

Conflict out of Teams

When you activate the "Help Process" using "safety teams," one important asset that flows from these group meetings is the "divergent views" of the members. If you are the leader of your facility, it is important that you "highlight" your expectation that this difference of opinion will exist from time to time and that you urge all views be added to the discussions that the safety teams have as they work to reach consensus.

In a regular, ordinary authoritarian organization, this safety management process and these "divergent views" more commonly surface as "conflict." The safety team members will need training in how to productively deal with the conflict so that conflict is a "strength," rather than a weakness. Conflict can be a "weakness" if there is no recognition that it is, first of all, okay to have a different view, and second, that the safety team has been taught how to deal with conflict, and to work out differences to produce the "best" solution.

Compromise is not what is meant. Frequently compromise is worse than either expressed view in a conflict. But what is derived out of this conflict is that the best parts of the conflicting views are brought together in a consensus where all agree that the team has reached the soundest decision on a safety issue.

In surfacing conflict in the work place, typically one finds that conflict seems to arise over who is supposed to do what. Such conflict points to an understanding of roles problem. Defining roles will frequently yield astonishing progress in work efficiency. This has been found to be a result of the productive redirection of energy previously spent in irresolvable role conflict situations.

Safety Role Definitions

Who is responsible to ensure that safety roles are developed? Those who lead must necessarily take the lead. Written safety roles contain expectations. Those who are led need to know from their leaders who is expected to do what. Where does my role on this subject, issue or problem stop and who do I pass the action to? Does this person know that the next action step belongs to them because they, like I, understand the roles?

I cannot emphasize too strongly the importance of safety role development. Most assuredly safety role definitions will prove to be a significant productivity improvement item.

Be very careful not to imply that safety role definitions are all inclusive. Understanding should be clear that role definitions do not contemplate complete job definition but cover the principle items only.

Where do I start with role development? What departments should I do first? Both are frequently asked questions.

There are three basic approaches to making such a decision. One is to start where the activity is highest and is most subject to short-notice changes that alter the workflow. This will be an especially lucrative starting place if many different people need to react to those changes.

The second is to listen for evidence of conflict. Where conflict exists there is most always an "understanding of roles" problem. Occasional conflict will arise from personality differences and we'll deal with these under "Working out Relationships" later.

The third is to determine the process by mapping or flow charting and then use this product as the sharing document as you "talk out roles." An important area in which to work out relationships for the safety team is to develop an understanding of the philosophical and real differences in the role of a Team Leader (the boss) and the team members. It will be different than just a "boss." Being a "safety team leader boss" is much harder.

First, the role statements must make it very clear that line managers, supervisors and foremen are in charge of workplace safety. They are supervising the work and the work process, thus they are also in charge of a safe work process and a safe result free of injury. Safety Staff are placed in an organization as skilled advisors, sometimes inspectors and auditors and frequently safety trainers. Safety Staff are, via safety training, inspections and audits, your in house safety compliance personnel.

Safety personnel must never become safety rule enforcers in a Zero Injury Culture. One cannot EVER punish one's way to Zero Injury. HOWEVER, neither can one tolerate continued safety non-compliance from any employee. Action must always be taken to assist non-performers into a mind-set of commitment. This is equally true of the management staff as it is the other employees.

Management staff should view themselves living in a "glass house." As such they must expect any move or statement they make to be seen and interpreted as a measure of their individual safety commitment. In a Zero Injury Culture ALL management MUST buy-in, period; and must reflect that buy-in without fail. Just as care must be given by managers regarding their spoken words, likewise, for a leader to omit verbal support of safety during a meeting is also a serious omission. If you are a manager and you "forget" during an opportunity to "speak out in support" of safety, then you need to re-examine your level of commitment to a zero injury safety culture.

Chapter 22

Managing the System

Avoid Temptation

There may be a temptation to shortcut what you have just read. My advice is to be careful. This is a proven system. Experience has shown that each Step of the system is vitally important to the health of the whole approach.

To be sure, as you manage the system after it has been installed, you will find it necessary to continually fine-tune each Step. New people will need to be "brought up to speed" and there is a very effective way to do this.

First explain the system and then let them become involved where appropriate; on teams, in the goals process, in the reviews. Recognize and accept that the reviews take time. It is very important to let those who were involved in establishing the goals to be a part of the progress reporting process and attend the goal review meetings with their supervisor and you.

Three benefits accrue. One, as leader you see each individual first hand and get a feel for their capabilities and potential; two, they can be a part of the presentation and thus they become much more committed, and three, they can hear your comments, concerns, and philosophy on a first hand basis. This gives them a sense of directed purpose without being autocratic.

In the bottom-up Goals process review, the Bosses' questions can easily become a part of the goal effort automatically due to the high interest level and buy-in to the bottom-up process. Care must be taken not to let your questions become unintended high priority goals. This tendency stems from a very basic fact. Committed people want to please!

When you see the system working this way, "automatic progress" is being made. Another way to see this phenomenon is, you now no longer have to ask someone to take on another challenge. You now need only to ask a question for goal clarification and any perceived added potential for progress is willingly added to the goal. And, just as importantly, if they are aware of problems that you did not perceive they will also willingly tell you about these.

Why? Because these are their goals! They own them and are asking for your buy-in in the review process.

Management Attitude

The attitude of Top Management is the single most important ingredient above all others that will contribute to the success of your "integrated management system".

First, support for the system must emanate from the top. Management's enthusiasm, dedication, along with your visible support, must pervade the management cadre as an expected norm.

Second, in a sense, Top Management will be living in a glass house. All eyes will be focused on these leaders; testing the resolve, checking for example. Questioning; "Do I see action that correlates with what I am hearing?"

You must pass this test!

Third, having worked through the steps, you will begin to experience the growth of satisfaction in your own mind as the "integrated management system" begins to work.

How to Make Changes

Every manager should realize that even though you develop this management system for automatic progress, on a random but regular basis events will occur that will create diversions. The need for additional goals that are short term and immediate will exist from time to time. Take these challenging events in stride. Fold them right into the management process. Priorities will have to change to accommodate the new goals but after all, the short-term need cannot be ignored.

On the longer time frame try with all your energy not to shock the long-term progress your system will afford you. If at all possible deal with the next needed changes in the work place during the next guidelines issue. This avoids diverting the energy that is on the verge of making that yet unseen step of progress.

So, you must be very careful in making any midstream changes in the guidelines. Direct the system as you would a symphony orchestra because in a sense you will be the orchestra leader and to change the score in the middle of the performance could easily destroy the quality of the finished product. Make your decision now that you want planned progress and realize to divert plans that are in place endangers all the work that has been done and serves to frustrate the people and the process. Decide now to consciously stand back and see dramatically reduced injury rates arrive!

PART 3

THE APPENDIX

EXPLANATION

THIS APPENDIX IS PROVIDED TO ALLOW IN-DEPTH ANALYSIS OF FOUR IMPORTANT SUBJECTS THAT THE AUTHOR CONSIDERS SALIENT TO ACHIEVING A ZERO INJURY WORKING CULTURE.

Item 1. "Sophisticated Safety Team Approach" Articles

The following articles are most suitable for a employment situation where the employees are in place for a five year to career length term. This does not mean that Safety Teams are not effective for shorter terms. For shorter-term employment situations the teams will not need to go to the depth of sophistication the articles contemplate. These articles are titled:

A. Team Management - What it is. What it is not.
B. Conflict and Teams
C. Team Decisions, Roles and Authority
D. "Openness and Trust" in Team Management

Team management is nothing new. Some leaders manage with teams without calling the process by that name. There are very sophisticated teams but "straight out of the box" teams are more abundant. These are very basic in their approach and for the purpose of obtaining safety management participation do not need a very high degree of sophistication.

If you wish to polish your safety teams to a high luster of professionalism then the following articles will assist that process.

The appendix includes four articles that were written "along the way" during the installation of the proven management system described in the book. They are written in the "here and now" time frame and had the common objective of helping the teams and people involved deal with some of the important interpersonal issues as they faced them.

These articles are offered in their original form and deal more with the function of a team in a general sense than just solely in the safety sense. They are for your reference and use as you set out to install "The System."

Item 2. The article on Monetary Incentives

E. Treatise on Monetary Incentives
In the decade pf the 1990's monetary incentives as a motivational approach to rewarding employees for achieving a zero injury workplace failed. Caution had been mentioned yet there are but few employers who still routinely give monetary incentive awards. What was the logic in support of these monetary incentives. Presented below in this section is an analysis of this body of logic.

Item 3. The Zero Injury Safety Techniques

F. The CII Zero Injury Safety Incentives Explained
The nine critical Zero Injury techniques are explained in this section of the appendix.

ITEM 1
The Sophisticated
Team Management Approach

A. TEAM MANAGEMENT. WHAT IT IS AND WHAT IT IS NOT.

Business Decisions

A business and the environment in which it exists is constantly changing. This change is caused by outside forces as well as forces from within and requires business decisions to be made in response. How well these decisions are made determines the current and future health of the business.

Research has shown that where practicable, decisions reached using a participative management style, where all information bearing on the problem is shared by those affected, are typically more sound decisions than those reached using other management styles. Or saying it in a slightly different way -that all the interdependent individuals, pooling all their knowledge as a Team, can more often reach a better decision than any single member working alone.

Managers installing team management need to make it clear that participation is wanted. A team member must understand that in accepting the role of involvement in the decision making discussion there is a trade off being made. For the right to participate in this team management process

206

one must vow to support those decisions reached even though the decision outcome may not go their way. This vow is in exchange for the opportunity to speak and participate and to contribute to and influence the decision making process.

When the members of a Team in a well-understood team process reach a decision, it is referred to as a "consensus" decision.

Team Management purpose is to reach the most sound business decision possible. The process provides for –
1. Subject understanding
2. Open communications
3. The development of trust
4. The welcoming of ideas
5. Sincerely considering all opinions

Thus, it is a "participative" management style that gives opportunity for involvement.

We must recognize that people differ in their opinions regarding most any subject one can name. It is this difference that generally ensures that all sides of a question are discussed. As a group discusses a topic with each advancing a view and presenting the related logic, a "body of opinion" forms where most of the group members are in "general agreement". It is at this point that the members representing the minority view have then to take a commitment to the group that they will support the decision completely, even though they differ. It is then that a consensus decision has been reached.

Without this individual commitment, the "openness and trust" advantage of the team style is lost when decisions become subverted by vocal dissenting team members. One may as well then forget "team" management and use the "authoritarian" style where views may or may not be solicited.

For emphasis, we say again that one exchanges the "secret dissenting" posture that frequently results when authoritarian management style is used for a chance to be heard and offer input to decisions. Thus one gives support to the team decision even on those decisions where the consensus may not totally agree with one's personal view.

Are Teams New?

In William G. Dyer's book, "Team Building: Issues and Alternatives", the first chapter is entitled: "Teams are everywhere." This is true. In reacting to the word "Team", we frequently think of sports. However, the Team concept applies to any group that has in common a "unity of purpose". This can be a chamber of commerce, a club, a family, or employees in a business where customer service is the objective. In this latter case, a restaurant is an example where all the employees have the single objective to serve an enjoyable meal to the customer. Including clean dishes, a well-set table, appetizingly prepared food and cheerful table service.

Likewise, as employees in individual functions or services, we work with our fellow employee to accomplish a given task. This task can be to operate a unit, to maintain equipment, to test products for quality, to provide materials, to provide advice, to keep track of costs, and the list goes on and on. All of us, by the nature of our work, are associated in some way with a group of employees who have "unity of purpose".

Actually, all people working in a manufacturing plant have a "unity of purpose" which is to manufacture products. Using possessive pronouns it would sound as follows:

"Our plant is ours to operate. Our Plant Team has this job to do. How well we do it depends very largely on how we relate to one another. If we relate poorly, we will not do a good job."

"The management style used here is one of the factors that determines how we relate to one another."

"It goes without saying, the problems are many that face us each day. How we approach these problems, how people work together in seeking solutions, reflects throughout the organization. If people work together and communicate well, the problems are understood and out on the table where we can see them and work on them; and we can move ahead. It is then, and only then, that a working climate can exist where all have a sense of worth and accomplishment. We can go home after our shift, or our day in the office, feeling that "today we got something done".

An Open Management Style

This "sense of worth and accomplishment" that we all like to have in the work place is enhanced by an open management style. Some call it a management style of involvement where each can speak their own opinion. Where all can be heard and each opinion does not fall on deaf ears; where each can call it like one sees it without fear, and where each can express their opinion freely in an honest and sincere way.

We need to recognize that with traditional management styles such an "all involved" working atmosphere has and can exist. This is especially so, where the leader of a group has a strong natural tendency toward the involvement style. There have been occasions where the leader invited all members of the group to a meeting and, in effect, a team session was held on a given problem.

Yet there is a distinct difference between a traditional management style and team style. Under most traditional systems, a leader works with group members as individuals. The most meaningful relationships exist between the leader and individuals. There is only limited interaction among the group as a whole. As a result, under traditional systems there is frequently a competitive relationship among the interdependent people needed to accomplish the task.

Team management reduces the non-productive competition among individuals working toward a common task. It brings these individuals together simultaneously with the leader and helps them understand each other. The result is an increased appreciation by the individuals, one for the other. This builds an environment of mutual trust, cooperation, and support, which allows better decisions to be made through the Teams.

Are there other uses for Teams? Yes, once teams are in place and meeting regularly, these meetings also provide excellent means to discuss and understand a wide range of subjects. They also offer opportunity to pass information through the organization. Thus, there is a considerable improvement in information flow and in the consistency of understanding.

However, team management, to be effective as a style, has to be the style people want. One cannot force a Team relationship to exist where one is not wanted. Sure, the team members can be designated, but this does not create a relationship of "openness and trust" that comes with mature team development. This is true at any management level in the organization. It takes cooperation one with another to have an open management style.

The act of "management" starts with each employee, hourly or staff. Each manages their contribution each day. Employees in the facility using their skills are in charge of and manage their contributions. A foreman, supervisor, manager, superintendent, or general manager is similar in this regard to a secretary, a clerk, an analyst, a draftsman, or an engineer.

Team Decisions - When?

A frequently asked question is "What kind of decisions can a "team" make? If I assist in decision making does this mean as a team member I have more authority? Let's explore situations where a Team is and is not and needed to reach a decision. Let's take the example of a car pool. Let's say a group of people get together, agree to share rides to work. They can discuss freely and seek agreement on: 1) what the driving rotation order will be, 2) what time they will leave the homes or work, 3) what route they will take, 4) how fast they drive, or 5) whether or not they will stop to have coffee or eat. All these subjects can be "teamed" because there is "unity of purpose" and all can contribute.

But if on the day you are driving to work, an eighteen-wheeler approaches head-on in your lane you won't need a team decision to hit the ditch; similarly, if "X" wants unleaded gasoline in his car, that's not a Team decision. Or if "Y" wants to use his own car rather than his wife's, that may be her decision and she is not on the car pool team.

So, one of the challenges in defining and using Team Management is that of understanding what is appropriate subject matter for a Team. Basically, it is much easier to "make team decisions" on longer range problems and procedures than those with immediate decision needs; i.e., the Eighteen Wheeler problem.

On the other hand, if the car pool team mentioned can get "Y" to agree to take up the subject of his wife's car with his "Family Team," then one Team can access another Team through the common member concept. "Y" is a member of the "Pool" Team and "Y's" own "Family Team". Also illustrated is that even though "Y" is on two different Teams, "Y" does not have more authority. However, the atmosphere does exist allowing "Y" to participate.

Similarly, in a team management environment it can be true that the leader of one Team is also a member of another Team. This allows communications up and down and across the organization to flow more rapidly, more smoothly and more effectively. With such a team management culture, one can surface concerns, long held back because an effective means to be heard did not exist.

We must recognize the distinctiveness of a team environment; it is where new ideas are encouraged and frank expression of views are not only allowed but wanted. Team meetings are the opportunities to air these views and ideas. As this team effort goes forward, also recognize that the day-to-day routine execution of work and the normal mode of overview and instruction given by supervision will continue to function. Altering the style of decision making to where certain selected problems are solved by a team does not change the fact that each still must manage their contribution and area of responsibility.

Team Relationships

One of the challenges is that of building this Team relationship. Successful Teams do not just "happen," the atmosphere of "openness and trust" must be mutually sought by all members. To do this, special Team Building meetings are helpful. These meetings are usually called "workshops" and are a day or two in length. Relationship techniques can be used by a facilitator to assist the Team develop as an effective work group.

A successful Team is very much dependent upon the leadership technique the leader uses. The leader must be willing to share some of the decision-making that he/she has control over with the Team. The key word here is "some" of the decisions. This does not mean "all the decisions" a leader must make can be put to the team. The leader decides which and when.

In this regard, the leader must not bring a subject to the Team for a decision unless the leader is completely committed to the concept of abiding by and living with the decision the Team reaches.

Further, the leader should not bring to the Team a problem where a decision has been reached at a higher management level and offer the problem "as a decision open for consensus" resolution. In such a case, however, the leader may be willing to give the Team the latitude to decide how to implement such a previous higher-level decision. The leader, therefore, has the important responsibility to identify these problems to the Team in a way that lets the Team know the decision constraints.

In a team meeting where a team decision is wanted, a leader must use care not to "bull-doze" personal ideas across. The leader's challenge is to maintain a role as a team member on an equal level with other team members while the various aspects of the topic are being discussed. The leader's role is critical in setting the climate so that the team can effectively reach a consensus decision on a subject that the Team is to decide.

Otherwise, team meetings will merely consist of the team

leader manipulating the team members to accept the leader's own strongly held views.

Team Management - What It Is - Is Not

Let's examine what "Team Management is, and what it is not" to a team member.

Team Management -

is not a way for each of us to get what we individually and sometimes perhaps selfishly "want" accomplished.

is a way for each of us to get what we individually want communicated to others. Also, to the degree that we can get others, through this communication technique, to agree; then, yes it is a way to accomplish what we want.

Team Management -

is not a style that yields permissiveness. The fact that a leader uses the team style does not remove the leader's authority or accountability for the results.

is how a team member, at the appropriate time, can raise an idea or issue regarding a rule, policy, or procedure and participate in the discussion. And "if" the team leader has delegated this particular area to the team, then a decision may be by the team.

Team Management -

is not the way to accomplish the urgent, immediate task. In these cases, we seek quick advice as needed and move rapidly.

is the method to get a procedure established that defines the way we make a decision in a particular urgent matter the next time it comes up. In such cases what the team is doing is this; if a particular identified event occurs in a sudden and urgent manner then the action taken will be thus, as pre-decided by the team.

Team Management -

is not a style that ensures that every individual will be involved in every decision reached.

is the way that the appropriate individuals are included in the decision process at the appropriate working level.

Team Management -

is not where a group of people get together and take a vote to reach a decision.

is where a group works to reach a consensus through a thorough discussion of all points and where all members voice their views; where difference in ideas are highlighted with the group seeks to take the best parts of all the ideas, putting them together to reach a better decision.

Team Management. Will it work?

The answer depends on our willingness to -
- Develop a personal attitude of openness and trust.
- Analyze what is appropriate for our Team to handle.
- Communicate our views in a style that does not criticize other approaches.
- Live with and support the Team decision even though at times it does not agree with one's own preference.

The team management approach is used to increase participation. There are many areas where experience will help the learning process. The challenge is to think through each step very thoroughly to ensure that each step taken is one that is positive. As progress is made forums need to be developed that allow individuals to share experiences with others. This allows a more rapid build in total experience base.

B. CONFLICT AND TEAMS

Recognizing Conflict

The news media of today typically uses the word conflict to reflect that a war is being fought or a battle is raging. For this reason, we sometimes hesitate to use the word conflict in a "Team Management" environment. Rather, we may say, "Why, we are a Team. Team Management means we have good relationships, we pull together to get the job done, and we seldom, if ever, have an argument. What do you mean conflict? That's a bad word around here!"

Yes and no!

Yes, we usually maintain good relationships, and yes we get the job done.

No, conflict is not a bad word around here.

The following pages will explain why "conflict" is not a bad word around here.

First, let's define "conflict" for our purposes in the management process. "Conflict" is simply a way to say "we are in disagreement. We do not see eye to eye. Our opinions differ on the issue."

Up front, let's set the record straight. "The better decisions are often born out of this reality of divergent views that we call conflict." Therefore, we need to learn how to deal with conflict and obtain the benefit from it.

Conflict resolution between party "A" and party "B" is one of four things. A wins B to their position, B wins A to their position or A and B each build on the ideas of the other to change a bit reaching a common position. Or after discussion, A and B find that the perceived difference did not exist. Research has shown that in solving a complex problem, two or more people (not to exceed a practical working size) putting their knowledge together using a productive process with a willingness to participate in give and take will usually reach more sound decisions than any one of the participants acting alone.

Many business firms go to great lengths to hire the brightest, the most innovative, and inventive people possible. It is people with ideas and with the ability to put their ideas to work in creative ways that yields progress. These people are typically somewhat aggressive. They are known as "doers". Let it be understood that one should not in any way diffuse this potential. Indeed, the opposite is true. It is this participative team management environment that ensures that this potential for creativity has a chance to surface and we can capitalize on this asset called conflict.

Our Values Examined

Okay then, with conflict out of the "closet", so to speak, let's think a bit about our personal behavior, attitude and "moldability" as we go through the process to resolve conflict. To some degree, this conflict we have comes out of differences in values from one individual to the next. These values are part of us. As we deal with conflict, we knowingly and sometimes unknowingly apply these personal values as we present our views and as we consider the views of others. Values come in different categories. Some of them could be categorized as follows:
- Moral
- Social
- Cultural

Some of our values are related to lessons we have learned during our lives that "taught" us how we feel and what we feel about certain aspects of living, working, managing, and interrelating with others. Our ability to "trust" others can be placed in this group if we recognize that moral, social, and cultural values also affect this "trusting" that we do.

There is, however, another source of values called "situation values". Sometimes our "situation" is referred to as "climate" or "environment" to describe our work "situation". If we set out to create a work situation that is enjoyable, we try to create an enjoyable atmosphere where people are friendly, courteous, and helpful.

Where we "value" these things we create such a place to work and then being friendly, courteous, and helpful become "situation values."

Right away, let's say we do not claim this list of "values" is necessarily complete. What we are saying is that this effort to list values, we feel, will lead us to better understand the reasons for conflict.

Our Judgments

Our judgments as individuals are based on our values, which on any given subject, will be fairly complex in the way the values are interwoven into the fabric of our opinion. If we are realistic, we must recognize that a really tough type of conflict that occasionally occurs is the basic "personality conflict." Such situations can be described as when two people due to personal traits that so annoy one another that they are unable of their own power to overcome their basic dislike for each other. It's sad but true.

It's also true that if we strive mightily, most of these can be overcome. Those that can't, must be recognized and the parties need to understand that in the final analysis our jobs do require these parties to work constructively together to accomplish mutually dependent tasks. While liking one another is nice and desirable in the work place, it is not mandatory. But I must say this; If general, wide-spread "dislike," one for the other, is pervasive in the workplace the organization will struggle continually. I also believe such a workplace is a very remote possibility.

We need to maintain that moral values and convictions are very important. In the business world, we must stand on a sound moral base of honesty and integrity to be successful in the long term.

Our Feelings

Added to the values we have, making them even more complex are our "feelings" about our values. Some of us wear our "feelings" up front, where we test what we hear or the actions we see and often react in a rapid, decisive, emotional way to what we see and hear. Usually, this type of reaction creates an instant issue. This kind of sensitivity is okay if we do not "turn off" the other party. It is important that we surface our "feelings" but try to be constructive in our response. A response that is so firm and unduly strong that it "turns off" the conflict process is not productive.

Usually, we will be working on conflict resolution where that conflict is based on opinion, feelings, and perceptions. And the parties must be sincere in their effort to arrive at the most sound decisions in the resolution of the issue. Also conflict frequently is caused by what one either "sees" or "hears" or in some cases what one "thinks" they see or hear. First, let's think about what we "see", then we'll deal with the problems arising out of what we "hear." In the realm of interpersonal relations where conflicts occur, the old adage still holds true that says:

"I can't hear what you're saying because of what I see you doing."

It's still said today that a parent who instructs a child to "do what I say, not what I do" is a very ineffective teacher. If we accept this as truth, then it is easy to understand that for "conflict" to have a chance of being resolved, what we say and what we do must be in harmony. Nothing is more defeating and discouraging than to think you have "talked" out a conflict and feel that understanding now exists and then discover the other party is still "acting" in the same old manner.

Now let's deal with what we "hear". Equally important in our "conflict" resolution process is our world of words; our vocabulary. We use words to express our ideas, concepts, and feelings. Our use of specific words can be the cause of an apparent conflict on an "idea" we are expressing. Not because of conflict in "ideas" between the two individuals, but rather because in the two minds there is a different meaning attached to a "word" that gets inserted into the conversation. All of us have no doubt in times past engaged in heated discussions arising from apparent conflict, only to discover that, as far as the real "issue" was concerned, we were not in disagreement at all and had suffered through the dialog due to personal differences in the meaning behind a specific word.

Our "Moldability"

As individuals, once we have what we say and what we do in harmony, it is then we can maintain awareness that there is the ever-present potential for ill-founded conflict to arise out of word usage. Our next challenge on the road to productive use of conflict is our "attitude" about the process. Our own "moldability", if you will. It is suggested that one of the secrets to getting the most out of the conflict resolution process is that of individual open-mindedness and personal "moldability". Moldable, in the sense that we as individuals are not suffering from a case of "tunnel vision." We are capable of seeing all sides of the issue. That we can indeed listen, hear, and understand from others all the factors they feel bear on the decision and are willing to listen. Yes, and even be persuaded to move or "remold" our position as we gain knowledge from the input of others.

At this point, we may find some people who are personally so competitive that their personality cannot tolerate "giving in". These persons tend to be immovable and as such are not very moldable. Hence, they have a basic problem that will prevent them from playing a full role in a participative management environment where the value of the decisions reached through conflict resolution is encouraged.

Our Patience

Now, one last point that is perhaps one of the most important of all is our personal "patience;" patience with the resolution process and patience with others. Opinions and feelings formed over years aren't changed instantly. It takes an in-depth discussion of alternatives and the logic behind them. It takes time to resolve conflict to get that "better" end product.

Summary

In summary, healthy conflict is productive. If we can effectively resolve our conflict and thereby capture the opportunity for progress, then our already good relation-ships can improve and we can get the job done even better and have fun doing so. One of the unique strengths of a team management environment that is lacking in other management styles is the opportunity it provides to more fully capture the progress that can arise from conflict.

222

C. TEAM DECISIONS, ROLES AND AUTHORITY

Safety Team Leader Role

Team Management as a name, implies a participative management style. In our routine use of the word "Team" it also implies a group with a leader, captain, or coach. An industrial setting incorporates a work groups' traditional "boss" as the Leader of a group as it seeks to become a Team.

A safety team cannot function without the willing participation of the Leader in the process. Actually, that is what participative management is all about: the Leader sharing his decision-making responsibilities with the group he supervises.

Borrowing from the above article "Team Management - What It is, What It Is Not", we find the following comment regarding the role of the Leader.

> "A successful Team is very much dependent upon the technique the Leader uses. A willingness must exist on the part of the leader to share some of the decision making with the Team.

> "In this regard, the Leader must not bring a subject to the Team for a decision unless the Leader is completely committed to the concept of abiding by and living with the decision the Team reaches.

> "Further, a leader must not bring to the Team a problem where a decision has been reached at a higher management level and offer the problem as one open for consensus decision. In such a case, however, the Team may be given the latitude to decide how to implement such a previously made decision. The Leader, therefore, has the important

223

responsibility to identify these problems to the Team in a way that lets the Team know what the decision constraints are.

"In a team meeting where a team decision is needed, a Leader must use care not to "bull-doze" personal ideas across. The challenge is to maintain a role as a team member on an equal level with other team members while topics are being discussed. The Leader's role is critical in setting the team climate so it can effectively reach consensus on items the Team is to decide."

Authority and Accountability

Let's examine then how one determines what appropriate matters are for a given Team to decide. It is obvious that to make a decision the Team needs the authority.

Authority, in corporations, is delegated down through the organization. Along with authority to decide, goes the additional concept of responsibility and accountability. These two concepts differ a bit from authority.

When a higher management level delegates "authority to decide," it has to release its right to make that decision. However, although higher management holds lower management responsible and accountable, it cannot thereby release its own share of the burden. Although shared by lower management levels, higher management remains responsible and accountable for any decisions made at lower levels in the organization.

It is the task of the Team Leader to decide the areas in which he has the decision authority and what decisions he will be willing to share with the Team.

Leader Sharing and Support

Remember, as teams are created, the decision process is moving from a basically authoritarian/consultative management style toward increased use of participative style. At each level, the leader is still responsible and accountable for satisfactory performance results in his area of supervision.

The leader will naturally have some apprehension about sharing the management role with the work group. As the leaders' confidence and trust in the work group as a decision-making team increase, an increasing amount of the decision-making will be delegated to the team.

There is a rather significant challenge facing the work group (Team) as well. As they find their Leader willing to share, the Team must realize that they are also responsible and accountable for the results of their decisions. The Team must, as individuals, be sensitive to the pressures that are on the Leader and give support and patience.

Likewise, the Leader must be alert to needs of the team members to be informed of the background information on issues that the Leader decides unilaterally or in consultation. Increased openness and trust are vital to increased use of the Team concept.

The Leader must learn how to share increasing amounts of information on decision issues with the Team. This will assist the Team in being understanding and patient toward the Leader.

A pitfall which some team members sometimes find themselves experiencing is the following: Typically, a team member feels increasing participation and opportunity to dialogue with the Team and Team Leader on many new subjects. As a result, the team member may tend to lose sight of their subordinate role and offer undue resistance when a Leader decides an issue without team input. At this point, the team style allows the subject to be raised and discussed at a team meeting, but it nevertheless remains the Leader's right to make the decision until willing to release the decision process on that issue to the team.

For instance, if on the issue of safety, a Team Leader felt the Team had insufficient aggressiveness in the pursuit of safety, the Leader may resort to consultative or authoritarian style in an attempt to raise the Teams' responsiveness to the situation. However, when one is looking for motivational techniques to deal with a subject like improving the Team aggressiveness in pursuit of safety, the Leader can use a combination of management styles.

The same (raising the Team's responsiveness to safety) task becomes more challenging if the person seeking improved aggressiveness in the pursuit of safety on the part of the team is a team member other than the Leader. In this latter case, the team member must use persuasive logic to win support from other team members for his position. In short, this member must do a "selling job". In such an example as this, the Team Leader faces the management challenge of openness in order to permit the interested team member to pursue this objective.

So one readily sees that the Team Leader sets the tone of team meetings, and through his "team leader" technique yields to the team in an appropriate fashion certain decisions. In all cases, it is the Team Leader's job to evaluate subject matter in order to assure the decisions that are made can be supported as sound alternatives. As a team matures in ability and understanding of its role, the Team can assist the Leader in this evaluation process.

D. THE CONCEPT OF "OPENNESS AND TRUST" IN TEAMS

Defining Openness and Trust

The words "openness and trust" are frequently used in discussing and understanding participative management. We hear them or see them and say to ourselves, "Oh yes, that's what we need alright. We need more Openness and Trust. I'll sure be happy when they tell me more and trust me more. That will sure improve how I feel about this place."

Wait. Let's not forget that openness and trust involves people; all people. For openness and trust to exist, the first step is to realize that it not only includes "they," it also cannot occur without "me." Each one of us must make our individual contributions for "openness and trust" to continue to develop.

To begin to understand the meaning of "openness and trust," let's deal with the word meanings separately and then bring them back together to see how they intertwine in concept.

Trust

Trust might be defined as the placing of confidence in another. Sounds simple. Why then do we not see more of it going on around us? Each of us can think of ways that we place confidence in others. One example is being acted out continuously on our highways, especially on those that are not divided highways. We drive our automobile on one direction at say 50 miles per hour while in the opposite direction a scant few feet away, a complete stranger does the same thing. We trust this stranger, and they us, to drive an automobile that is in safe condition in an alert and responsible manner.

A slightly different situation is experienced when one boards an airliner, or a bus or simply agrees to ride while another drives.

Thinking in this way we can literally think of thousands of illustrations of "trust" in our society.

If this much trust is going on within us and around us, why then suggest we would benefit from "more trust." These things we think of are examples of trust that are exhibited in physical ways. Another area in which we experience "trusting" is in the emotional, ideological, and personal information areas. These areas are "what we are," "who we are" and "how we feel."

Trusting others to perform in the illustrated physical way is somewhat easier than trusting another with information from our personal knowledge base about our work or about our selves. Especially so, if this is personal information which could damage us were it handled in an irresponsible and non-confidential fashion. The common question then is "Is it being suggested then that you want me to divulge more personal information?

We ask, "For what reason?"

We reply, "That information is indeed personal, it is mine, mine alone and I do not want it known, you can forget it."

Right on! Divulging that personal information that you do not want known is not required in teams. We are talking primarily about sharing work related information in a more effective way.

So now we are into "divulging" information to someone we

trust. Isn't that the same as "openness"? Yes it is and thus the first evidence of the inter-dependency of the words openness and trust; but on with "trust."

Consider the following statement: "For me to trust you, I have to feel you are trustworthy. That if I become more open, you will handle that openness in a responsible fashion. That you will not joke around with that which is serious to me and that you will not use this bit of information against me, or to take advantage of me."

Developing Trust

Okay then, if that illustrates the "trust" we are talking about then how does one go about developing trust in relationships on the job?

First, we must communicate. Communicate, with each other, something of ourselves, our ideas, our concepts, in the areas of what we are. Both in terms of our job and related information and in terms of emotions, ideology and some amount of information that helps the other to see us better, to understand our motives.

Second, we must develop a willingness to accept that which we have learned about the other and at the same time also accept the person. Each person possesses a specific set of personal values. These values are measuring rods. We use them to evaluate situations, what we hear, see and yes, other people. In a sense "I am what I am" and "You are what you are" and "I accept you." "I may or may not agree, but I nonetheless accept you."

Third, along with being accepted by others we need to improve our ability to better understand ourselves by learning to look inward at what we really are. This does not at the same time imply that we are necessarily "happy" with ourselves. It may be that we may want to change a bit to be more effective in the area of human relationships. This change we may want to make can be in one or more of different specific ways. An example might be that one would want to develop their "listening" skills, or perhaps to be less "impulsive" etc.

Development of trust in relationships is not a rapid process. For instance we meet a person, we begin to talk at first about our hobbies, our families and about our feelings. It is at this "feelings" level that we begin to develop a understanding about this person and they us. The sharing has already begun. The opening of self ever so slightly has begun, always at first with those little things that are part of us, but that are easier to share and are more difficult to misuse. Then later, if the mutual interest allows and mutual respect builds, a bit more information of a slightly more risky nature is divulged and so on. This process is rarely a rapid one but usually takes considerable amounts of time.

This revealing of self will form a small segment of the typical information flow that is discussed in the workplace. The majority being the information we have as a function of our job. Included in this will be information that comes as part of our job objectives, our job role, our job authority and our personal information about our job.

This job related information bank is what we use to do our part in contributing to the work accomplishment process in our work place. Our communications to others from this information bank comes in both verbal and written forms. Sometimes it will be a decision that is directional in nature and we pass it on as "what must be done." Sometimes it will be simply a small segment of information that another person or group can use in their work process or decision making.

This information that we pass on is our individual contribution. Others see this information that we offer out of context with our personalities unless there is a mechanism and a willingness to let ourselves, our motives, our thinking, be known. As we develop "Openness" in this way, it relates to our work or job oriented information and then personal openness begins to have a direct impact on how effectively we work together in getting our work done.

Openness

Some people are not naturally very open or trusting tend to share minimum amounts of information. They tend to have just enough openness to contribute to a decision but no more. Often background information that reveals the "why" behind the information offered will be omitted. In such cases the decision gets made but people do not understand why and have minimal sense of participation.

As we "open up" a little, higher levels of understanding are created where eventually a person can know the basic information behind a decision and understand it more fully. At this point perhaps an optimum level of knowledge exists and people feel informed and feel that they are contributing from a well-balanced knowledge base and have a feeling of importance in their job.

Participative or team management carries this "openness and trust" concept a step further. In appropriate situations the objective is to allow more of the decision making to drop to lower levels in the organization. Along with this decision making authority of course must flow all the necessary information from across the organization to the decision-making "Individual or Team."

With this information the deliberative process can begin. In a team discussion, each team member offers their own

231

appropriate contribution, as a function of their knowledge base. Each team member strives to be trusting, communicative, and accepting of others in order for all members to have an identity with their personal values known and understood. This "openness and trust" then has a direct influence on the quality of the ultimate decision and each feels free to contribute.

Openness and Trust

"Openness and trust", two words intertwined inseparably in a participative work place contribute significantly to the creation of an effective working environment.

What does an organization gain from such development? The simple, but true hypothesis is that as we know one another better, appreciation for one another develops, anxieties and uncertainty are reduced, and the organization functions in a much improved fashion. Last, but not least, such an environment permits considerably more enjoyment to be realized as our work is performed.

ITEM 2

Treatise on Monetary Incentives
Re-wtitten in 2011

The Arguments:

Monetary Safety Incentives for the worker?

Please note that my position has firmed in opposition to the use of monetary incentives for reasons explained above, in order to be balanced in the treatment of incentives I am including this subject in the appendix. All readers likely know of OSHA's investigations into the use of incentives since incentives are alleged to increase the number of unreported injuries.

Safety Awards versus Safety Incentives

It is important to differentiate between "safety recognition awards" and "safety incentives." I cannot really say at what value an award becomes an incentive but the concept is that if an award gets large enough it becomes an incentive. Or if an award is known in advance it can become an incentive.

The opponents to their use define an incentive as when the value of the item awarded exceeds the "monetary" value that causes the first worker to fails to report the award becomes an incentive. In other words, as an incentive increases in value at some point it is likely to cause under-reporting of injuries then at this point it becomes a deterrent rather than an incentive. There is logic to this argument.

There is a "Vested Interest Concept"

It is quite rational thinking to expect an individual to protect that in which a vested interest is held. In this regard, each of us works to keep alert so we protect our body against the pain and suffering caused by injury. Therefore, one can expect a worker to routinely do his or her job safely, avoiding injury to self, insofar as knowledge of danger and absence of a reckless nature will allow.

This instinct for "Self Protection" is obtained through personal experience and from the knowledge of others (teachers, trainers, instructors, supervisors and other colleagues) and their written materials. "At-risk behavior" is can be brought under control through the same educational means.

Through safety education our "vested interest" level is empowered thus raised to where our "self preservation instinct" comes into sharper focus. We learn to think before we act, or as is the case many times, we learn to honor our cautious thoughts more readily. Importantly all should realize that safety training equips employees to be aware of "at-risk" behaviors the employer wants them to avoid.

The Paramount Question

Even so, a paramount question arises in worker attention to personal safety. "Can a worker's level of safety awareness, and thus their 'vested interest' and `self preservation' instinct be stimulated in a positive manner to higher than routinely normal levels?" It is quite common to see or hear in the testimony of an injured worker the expression "I knew I shouldn't do that," or "I just wasn't thinking."

Hearing these testimonies the question again arises. "Can one's self-preservation instinct and thinking be stimulated by means outside the individual to greater levels with the result being to raise to a higher level the instinct of self preservation?" And thus achieve a higher level of attention to "injury avoidance."

Some Use Safety Incentives to Do This

Safety awards are recognized as gift items that are used by employers to show appreciation for safety achievements.

Do they increase worker attention to "injury avoidance?"

Combinations of awards such as belt buckles, ball caps, tee shirts, jackets, decals, gift certificates, dinners, picnics, and other "things" have been used for many years. Many feel that these awards are useful in showing appreciation and thus stimulating employees to believe the employer really cares for them thus causing an increased focus on avoiding at-risk behavior?

Others argue that while showing appreciation is very important and employees do feel good about receiving a gift, mere recognition per se' does not stimulate a safer worker unless the award takes on a significant value such as a monetary incentive.

Prior to making the decision to use monetary incentives there are some important "first things first" considerations that an employer should note.

The Importance of a Safety Program

A well documented, publicized and worker understood "safety program" that includes the CII zero injury research results must always be the foundation on which any safety incentive programs is managed. Once the employer has the "safety program" in place and safety orientation and training ongoing, what else can be done to further lower the occurrence of injury?

How can a company's employees routinely achieve the level of "Zero Recordable" injury?

Can a worker's "self preservation" instinct be raised to higher levels through monetary incentives?

It seems to most employers that it is impossible to get beyond the continued occurrence of that "it should not have happened" injury." Yet, there are those who do achieve "Zero Recordable" injury for extended numbers of hours worked. And they accomplish this without the use of monetary incentives.

Author's Changed Opinion

I will enter my personal observation here after over 20 years working in the field of zero injury while advocating the use of the CII Zero Accident research results. I admit that in 1993 when the Zero Accidents Task Force reported out their research results there was strong evidence that those contractors who were paying monetary incentives to their workers routinely had fewer reported injuries and I, as task force chair based on this evidence supported the use of monetary incentives in the form of cents per hour wage incentive if the project was injury free.

It was in the 1995-1999 time frame out on project sites interviewing crafts that I found that there were many crafts that were against incentives for they were seeing first hand that injuries were going unreported. After actually interviewing craftsmen who were personally harboring an unreported injury in order to receive the incentive I admit I had to change my opinion. It was clear that injuries were going unreported and it was also clear that as a result, the so-called zero injury work site was violating the honesty and integrity values all contractors wanted their employees to observe.

The many contractors who tried incentives also found the same result and soon all but a "few" ("none" that I know of) ceased using the safety incentive option.

Personally I am now at the point I do not even like to hear the word "incentive" used in conjunction with the safety program always using care to use the words "recognition and rewards." But I also have to admit the word "incentive" is nonetheless routinely used on projects to describe "recognition and rewards." And to some degree the use is correct however the items used as gifts are small in value and in a properly applied zero injury training program are mentioned to the employees undergoing safety training only as evidence of "recognition and appreciation" of a job well done.

The critical aspect is this; and it will forever be with those trying to show the essential appreciation of good work. If a recognition item surpasses the moral threshold of what it takes for the first worker to not report an injury in order to receive the appreciation gift item, then all have to admit we have a situation where an appreciation reward just became an unwanted incentive.

Sum it up: I am strong for showing appreciation! I am strong for doing so based on leading indicators as opposed to the lagging indicator of a recordable rate.

I am firmly opposed to any appreciation process that will cause an employee to not report an injury in order to receive an appreciation gift item.

Here I advocate innovation. As you innovate be assured the subject is very complicated because it gets into the area of motivation of employees and motivation of employees is all about winning the mind of the employees to expend extra-ordinary energy in avoiding at-risk behavior. To not report an injury is just as surely at-risk behavior as is a safety rule violation and should be treated as such. So it becomes clear if an employer is not very careful the employer through trying to do the right thing becomes guilty of causing that which they do not want; unreported injuries.

The "Zero Injury" Achievers Are Increasing In Number!

Contractors who have embraced the notion that Zero Injury is a workplace value they want their companies to adopt have steadily increased over the past decade. These have implemented the research based results known as the Construction Industry Institute Zero Injury techniques.

Many have also tried the Construction Industry Institute 1993 reported Zero Injury Technique of "monetary incentives." Some have used this feature in conjunction with other awards of lesser value while a few others rely solely on the monetary incentive. These are those who do feel the incentive feature assists them reaching that illusive "Zero" injury level.

The truth can only be determined by a third party conducting confidential interviews in what I term a "safety culture assessment."

Some have tried monetary incentives without the CII Zero Injury Techniques only to find no success. As the research indicates, success still requires that most important ingredient of all, "skill in human relations" and use of the Zero Injury Techniques.

Selling employees on the concept of Zero Injury requires a sincerity of purpose and a "caring" that is recognizable by the employees. Employees then become co-advocates of a Zero Injury working culture, the result being Zero Recordables for the life of the project or as a minimum, for very long periods of time.

Success with Incentives or Deception

A "monetary incentive" is a "cents per hour" bonus for safety excellence. The proponents testify they are reaching "Zero" injury with increasing frequency. The money is paid in a separate check given to the worker by the foremen, or higher authority at the end of each injury free pay period or at a minimum, monthly. The incentive is "always" for all employees or none and is based on the performance of all. All employees, including foremen, receive the bonus if there are "Zero recordable" injuries for the month. If any employee suffers an injury during the month no one receives the bonus. In the past some employers have given the incentive for a "Zero Lost Workday Case" frequency. I know of no one doing so in 2011.

The Arguments Against

There are four recognized arguments against the use of monetary incentives.

These are:
a. "where does the bonus money come from?"
b. "why pay for self preservation?"
c. tempted to overuse "light duty" and
d. the working injured problem:

Relationships Are Important

The relationship among employees and the relationship between employees and their supervision is a significant factor in safety performance. If, through common vested interest, one can open up these lines of communication so information flows more openly and freely, then a zero injury project is more easily achieved.

On the other hand, if employees do not get along well with each other and if employees hold their supervision in disdain, then the opportunity for zero injury to be achieved is severely impaired. As it always has been, "Success in safety excellence is highly dependent on mental attitude and mental attitude is dependent on relationships."

Incentive proponents argue that collectively awarded monetary incentives improve relationships and open up the channels of communication. And when employers care enough to give monetary incentives, the supervision (also getting the same incentive) are motivated to keep the communications open by treating the employees with an increased level of concern.

They all have a common "vested interest" in achieving a no injury job-site. The "self preservation" instinct is increased and the quality performance level of Zero Injury is more easily achieved.

Back to the Arguments Against

Let's now go back to the arguments against the "safety incentive bonus" approach and take them one at a time, last first.

A. Where does the safety bonus money come from?,
B. Why pay for self-preservation?,
C. Tempted to overuse light duty, and,
D. The working injured problem.

The Working Injured

The working injured concern results from the temptation of a worker to hide an injury and continue to work with possible aggravation of the original injury. This is a real issue and in a monetary incentive environment steps must be taken by the employer to ensure that this does not occur.

It is a fact that employees have been known to hide injuries because they did not want their colleagues to lose their safety incentive. OSHA looks skeptically at monetary incentive programs if they are administered in absence of any real substantive effort to create a Zero Injury culture.

We all know that occasional cases of non-reporting of injuries occur on all projects. Monetary incentive opponents believe incentive increases the chance of this occurring. But, proponents say to use this reason as a principal argument against safety incentive payments may be to say, that eliminating the human suffering of work place injury to at, or near, zero injury through improved safety communication is not worth taking the chance that a few employees may aggravate a hidden injury.

Overuse of Light-Duty Work

Proponents believe light-duty work properly assigned is equally productive to any other work. Non-legitimate use of light duty-work is not appropriate in a Zero Injury management effort.

With reduced injury frequency will come a reduction in the need for light-duty assignments. If a light-duty policy is seen by the workers as a sham then likely light-duty is being overused or at a minimum miss-used.

Paying For Self-Preservation

Proponents argue that a safety bonus is but a "productivity payment." In this sense they maintain it is no different than the base wage. Worker productivity is exchanged for base wage. Separating safety performance, an easily measured metric and productivity item, from the base wage concept and using a bonus for zero injury achievement is as sound a concept as base wage they argue.

Where does the incentive money come from?

The money we are talking about is now being spent for injury medical expense! The direct and indirect cost of worker injury is much higher than most people realize. Though some of the cost is at the expense of the injured worker; i.e., reduced wages and the inconvenience of injury, the larger share of the cost falls on the employer. Injury costs are much higher than prevention costs!

Research has shown that the indirect costs of injury exceed the direct costs (those costs covered by Workers' Compensation insurance) by a factor of 2 to 20 to one. Just looking at base Workers' Compensation premium alone one finds the cost of construction worker injury as determined by the premiums paid for insurance ranging from $1.00 to over $5.00 per hour depending on the trade across the 50 states with the national average being around $2.00 to $5.00 per work hour in higher paid trades.

Reducing injury in a company to levels of near zero saves a large amount of money that historically has reduced an employer's ability to compete and has directly reduced profit margin. Proponents of monetary incentives believe that sharing a portion of the savings accruing as a result of safe work with the safe worker will create a "common vested interest" climate. The employees get an incentive and the employer enjoys an improved profit margin.

Proponents argue that such proves that management is willing to put their money where their mouth is!

Incentives Summary

There are a number of serious issues that have to be managed if one chooses to use monetary safety incentives. One of these is the tendency for the monetary incentives to become "owned" by the employees as part of their wage. All effort should be utilized to avoid this possibility. Were a company to install monetary incentives company-wide on a basis that was not very difficult to attain, and if they kept these in place for even a short period of time the fact is the incentives will become part of the working conditions and can easily become viewed as owned by the employees. In a union represented workforce this can become a bargaining issue.

Equity, employee to employee, becomes an issue in the administration of any incentive program that is not totally equal in how it rewards employees. Equity is an employee morale issue thus must be watched and managed to minimize this problem.

My Personal View

Being against incentives what can what advice can I offer nonetheless to those employers using them. It is mandatory that any incentive programs be preceded by an aggressive drive to bring the safety program into compliance with world class safety technique content which is defined by the CII zero injury research results.

Otherwise a safety incentive program based on the lagging indicator of worker injury will inevitably drive some injury reporting underground. After all, employees do want the incentive money and some will be tempted, and afew succumb to getting it by not reporting injuries.

On more than one occasion I have heard of cases where employees have gone to their own doctor after hours and paid out of their own pocket the costs of an injury rather than reporting an injury and causing all their working colleagues to lose the safety incentive.

Counterbalancing this, I have also heard of employees who were injured at home but brought the claim to work and alleged the injury occurred at work. In such cases incentives will curtail this inclination due to the same peer pressure that causes injury to be underreported.

But on balance these under-reporting issues give rise for me to caution "would be incentive embracers" to wait until your zero injury safety culture is in place. Then if used, try incentives only on projects where it can more easily be managed with safety personnel on-site to assist in ensuring injuries get reported.

Some Achieve Zero Injury through Extraordinary Skill

In support of those opposing monetary incentives there is something to be said for those who achieve Zero Recordables with pure safety leadership skill.

Getting to Zero Injury through years of dedication of resources and the application of high levels of people management skills have always occurred in some companies. Most of these have not used monetary incentives. However they do involve their employees in the management of workplace safety and routinely recognize them for their achievements. There are major and some minor corporations that have been able to achieve the "Zero Recordable" injury level unknown to the rest of us. When you think about it\, it was some of these companies that were found to be working injury free with this discovery resulting in CII mounting its' original research efforts to answer the question of "How do they dol that?"

Extraordinary skill in human relations is always found to be a vital ingredient in all zero injury achievements. If the worker truly feels a part of the safety effort of the corporation, the worker's "vested interest" level increases and safer work occurs. Interestingly, another thing that is "always" present in these companies is a passionate zeal for safety on the part of the Company Owner or the Chief Executive Officer.

ITEM 3

The Zero Injury Safety Techniques Explained

1. Demonstrated Management Commitment –
Restated from Chapter 5

The key research finding was:

In the zero recordable injury companies the CEO always had a key operating safety expectation placed before the company management.

It is paraphrased as follows -

> *"It is my desire that we will do our work without injury to our employees. My belief is that all injury can be prevented and it is my expectation that there be no worker injury in our facilities or on our projects. If an injury does occur it will not be viewed by me as an acceptable event! And I personally will be involved in determining in what ways management failed to allow the injury to occur.*
>
> *We will not set goals for injury! Our commitment is to ZERO Injury! This is not to be a statistics management effort. Rather our commitment shall be a complete devotion to the elimination of situations where our employees are at-risk or through education bring our employees to realize that any unsafe behavior is not solicited nor*

desired. I urge all employees, management and workers alike to willingly become partners in our commitment to eliminate employee injury."

In Chapter 16, I offer a prioritized plan that leads off with management taking action. Further, in the 2001 CII research, the task force developed additional measures. These are shown below. These give specifics on how to ensure that top management's participation is routinely applied in a productive manner.

With each CII technique the Recordable Incident Rate results for the contractors using, versus those not using these techniques is given.

These are:

1. Top management participates in investigation of Recordable injuries.
- Participates in every injury – RIR = 1.20
- Participates in 50% or less – RIR = 6.89

2. Company President/senior management reviews safety performance record.
- Yes – RIR = 0.97
- No – RIR = 6.89

3. Frequency of home office safety inspections on the project.
- Weekly/Bi-weekly – RIR = 1.33
- Monthly/Annually – RIR = 2.63

Creating a Zero Injury work place culture is an initiative that must be led by top management. To delegate this all-important activity to lower levels of management sends a signal that such a culture is not that important. With that signal the implementation effort is severely hampered.

2. Safety Staffing

There is a battery of information within the construction

industry from those that are successful in achieving the Zero Injury performance which indicates that ratios of employees to Safety Specialists should not exceed 1 per 100. The CII 2001 research reported that ratios of 50 to one is common on large projects where zero injury is the performance norm.

It will be no surprise if some in management wonder how a company can afford such a heavy contingent of safety professionals and specialists?

The answer is in two parts:

The first is: Many managers who are successful in achieving Zero Injury performance find such performance brings with it so much more than just a safety result that they cannot see the ratio being otherwise. They feel this way because they see the profound connection between safety excellence and reduced cost/schedule that they are convinced that operating in the Zero Injury culture produces the highest return on the investment. In these companies safety is of paramount importance and the staff is required to be in compliance with all company and regulatory requirements, and to guarantee that people are available to ensure the highest quality in employee orientation and training.

The second is: The proper ratio for each company is obviously heavily dependent on the nature of the work each employer is doing. The 50 to one ratio was found in the large construction projects and large construction firms where employee turnover is very high. Smaller firms in the construction arena that have less turnover will require less safety staff with firms that experience little employee turnover having the highest ratios.

Remember this, the proper ratio is merely that required to ensure all personnel receive quality in safety orientation and in safety training. Add to this the requirements for OSHA compliance, safety inspections and audits, the general maintenance of safety processes and procedures and you have near the correct number of safety personnel.

It is a given that the safety personnel you do have must be sold on the Zero Injury process and devote their creative skills to developing quality training and orientation materials for the employees.

3. Safety Training

Pre-Project/Work Safety Planning

"Weeks" before the start of a new production effort or project, the leaders, including any contractors get together to determine and discuss the various safety hazards the parties will encounter as the work progresses. The word "weeks" is used to reflect the need to allow ample advanced planning in order that arrangements and coordination time is allowed for the parties to arrange for equipment and special tools to be acquired prior to the work requiring them. The point is that the Pre-Project Hazard analysis be held sufficiently in advance to allow for any planning that might result to be executed in a timely manner.

Attendance at this meeting should include all affected parties including Safety Representatives from any participating contractors and sub-contractors. Ample time should be allowed to go over all the aspects of the work to be done from beginning to end. Ample time for the meeting should be allocated to avoid appropriate safety planning from being deferred to a later date, due to the lack of time.

Pre-Task Safety Planning

Pre-task safety planning is a safety tool to be used by the foremen and the crafts. It is not difficult to do. It is so simple in approach that the biggest threat is that your people will think it a trivial matter. Do not be deceived. Pre-Task Safety Planning is a VITAL ingredient to achieving a Zero Injury working culture.

The most successful use it formally through the design of a form to be used by the foremen each and every time a task assignment is changed or given, and it is never used on a lesser frequency than each morning.

Typically those that use forms make them into a simple checklist that serves as reminders of the "at-risk" behaviors to be avoided on the tasks and to make work plans for task execution. When foremen and crew make these "pre-task plans", they serve as coordination for the execution of the work.

Those who use the pre-task planning process warn that if during the execution of the work the execution sequence is altered, stop and make a new plan. For, they say, injuries frequently occur when changes are made without proper input and knowledge of all involved. It is OK to change the plan. Just do so formally. Stop and make a new plan.

The testimony of those who use the pre-task planning process is that their crews are more productive, leading to work completed ahead of schedule and on or under budget. These two aspects of Pre-Task Safety Planning are the reasons you will want to ensure complete and total utilization of the technique.

4. Safety Training and Education

Safety Orientation

Safety Orientation of all new employees before they commence work is the objective. After all have been given the orientation covering the logic behind the Zero Injury commitment on the part of management, the effort then becomes centered on the new employee. Each new employee receives safety orientation prior to reporting to his/her foreman for a work assignment. It is during this orientation that the new employee learns about what it means to work in a culture where Zero Injury is the expected and supported norm of operation. Thorough safety indoctrination on the zero injury safety culture of the employer is the objective.

Know that each new employee comes to you as one who may have never heard of anything like what you are working to achieve: the completion of an entire year or project without an injury to any employee. Therefore the first thing that has to happen in orientation is this new way of thinking and working has to be explained in-depth ending with the new employee being asked to become an employee committed to working safe.

Defining what Zero Injury means in worker terms and the logic behind this concept must be presented. Doing this will help to "sell" the new employee on being a committed and participating member of the zero injury effort.

Also it will be appropriate to show the worker the various safety videos that can be a part of the new employee orientation. Videos on tool usage safety, behavior expectations, etc., are in order. But never rely just on videos for proper orientation. The number of employees attending is not critical as long as the facility is comfortable and the orientation session is varied to maximize the attendee's chances of absorbing the information.

In the ideal construction scenario the Orientation should open with the Owner Representative and the Project Manager giving a few brief remarks about their own commitment to safety. In a manufacturing situation it is desired that a member of management open the session. Then in either case the orientation presentation is turned over to the instructor designated to give the orientation. This can be anyone who has a passion for zero injury and basic presentation skills. Never allow an orientation to become a ho-hum process that lulls the new employee/s into a state of boredom.

Many companies use Safety Personnel to conduct the orientation, others use line management personnel. While the person used can be largely a function of availability, never use someone who is not gifted in the art of presentation and can display a passion about the subject of employee safety.

Time allocated for new employee Zero Injury orientation across industry varies from one to eight hours. My idea is that more than one hour is appropriate and more than a passing effort to provide the new employee all the information needed to become a participating member of the zero injury effort must be made.

The sessions can also include safety training in critical areas where appropriate. Among these could be the use of fall protection gear, respirators, and certain special tools with the latter being more trade specific.

Once again be careful to put all foremen through the same safety orientation that the crafts go through. The other option, which I have found in more than one instance, is not a pretty thing to contemplate. That would be sending new employees out to their foremen without the foreman having a clue as to what they have been told in orientation. Despite the obvious nature of the need to give foremen orientation it still happens that I find facilities and projects where the foremen have inadvertently been omitted from orientation.

Safety Training

In the 1993 CII research, when the researchers asked the crafts people what they thought was most important in achieving zero injury they answered with the following six primary items.
1. Fall Protection
2. Pre-Task Safety Planning
3. Safety Person on the job
4. Safety Training
5. Protective Training
6. Recognition and Incentives

Notice that two of these items are on training. The crafts personnel wanted Safety Training and specifically training on Personal Protective Equipment. Some of this can come during the safety orientation before reporting for work. Other training should come throughout an employee's time with the company. Today many companies give their employees training in OSHA compliance, CPR and First Aid training. Some even give an OSHA 10 Hour course. In the research when the researchers asked "How much time do you spend in Training?" the answer averaged 15 hours per year per employee. This is equivalent to two eight-hour shifts.

5. Worker participation and involvement

There are a number of methods available that allow the critical worker participation in the effort to achieve a Zero Injury culture. The three most used are safety team membership, behavior based safety observations, and participation in safety audits and inspections.

The last is likely more suitable to rapid turnover workforces while the first is more suitable to situations where turnover is rather low.

6. Recognition and Rewards

This is a very critical item and many do not give it enough attention. For your employees to remain loyal to the Zero Injury effort it is very important to routinely give them recognition for worthy achievements in avoiding injury.

Typically an employer is looking for significant improvement. It is critical that as this improvement occurs that the employer have ready an appropriate safety award with which to recognize a safety performance milestone. Obviously if an award is to be given soon, as in immediately upon achieving the milestone, some planning and award acquisition has to take place before the milestone is reached. I urge you to accomplish this preparation in a timely manner.

It is very discouraging for the employees to know they have achieved a laudable and praiseworthy milestone, only to see management remain silent and as far as they can see inactive in preparing for recognition activity.

Remember research has found that "praise" is ranks higher than wage in motivating employees to excel.

Some employers find it appropriate to recognize the absence of Zero Lost Workday Cases as the first milestone then possibly move this recognition to Zero Recordables as the performance continues to improve.

Regarding the question of worthiness; when is an achievement in eliminating injury worthy of recognition?

Here I strongly urge employers to keep historical data and use the national averages in the appropriate industry as your guide as to when to make an award. One easy milestone and usually a difficult one to achieve is when the employees have worked 200,000 hours with zero recordable injuries.

This is equivalent to 100 employees working one year without an injury. That is the definition of Zero Injury when considering OSHA Recordables. On the BLS/OSHA base of 200,000 hours the achievement of zero recordables is the objective.

After this the objective becomes a repeat of the same achievement as many times as you can. The record as far as I know in repeating this 200,000-hour record back-to-back event is over 23 times in a period of 14 months in the construction industry. That is equivalent to over 4,600,000 recordable free hours worked. Based on the 2003 BLS national average (when this record occurred) this performance avoided some 156 recordable injuries which is amazing!

Caution; do take a look at your industry and avoid making awards for safety achievements that are in fact mediocre or slightly better than average. Remember you are looking for zero for significant hours worked, so be careful to give your awards based on this objective.

7. Contractor/Subcontractor management

This item applies to General Contractor's management of sub-contractors. However, the same emphasis is made for facilities in managing the contractor workforce that you retain to work in or on your facility.

Safety management usually begins with your own workforce. Soon after implementing the zero injury effort for your own direct hire employees it is timely to begin to work with the contractors who work for you to steer them toward the zero injury concept. The aim is that over time you will only retain contractors to work for you that have the same devotion to the zero injury effort as you do.

There are some legal implications and these can easily be avoided. I am not an attorney so do check with your legal advisor here. Typically a purchaser of a contractor's services cannot legally and should not be directly stipulating how the contractors you "have retained" should manage their safety program.

However the method that is typically legal in most states and municipalities for private work is to set the standards contractors have to meet to be considered as potential bidders on your work before you retain them. In the area of safety you simply state what the safety program requirements and content are for a contractor to qualify to bid your work. If the work is for a government body there is often an open bidding requirement so this will govern. If possible encourage the body seeking the bids to make some requirement in the area of safety. Such an effort is sometimes politically challenging and takes time.

Another aspect of working with a number of other contractors where the safety performance of any one impacts others is that it is highly productive to hold contractor's safety coordination meetings at least weekly but more often in highly congested areas.

8. Accident/Incident Reporting and Investigation

There are four parts of this area,
1. Tending to the injured,
2. Reporting the injury
3. Investigating the injury and 4. Managing the injury.

1. Tending to the injured.

Primary in any injury situation is the proper care of the injured employee. The foreman must immediately arrange appropriate medical care as dictated by the extent of the injury. This can include the foreman transporting the employee to treatment/ medical care. If another supervisor or the safety person is available, these may assist or in some cases take over the care-giving role.

2. Reporting the injury.

Part of a Zero Injury culture is the specific accountability of "line management" to act as the primary means of insuring proper medical care is given as well as the communication link to top management when injury occurs. The reporting begins immediately with the foreman reporting to the second supervisory level.

The second level supervisor must be informed as soon as practical. First priority is always giving primary attention to tending the injured employee. This initial report may be delegated to a peer employee or another foreman in the area. Upon being informed the second level supervisor's immediate obligation is also to the injured to ensure that all appropriate resources are made available to ensure rapid care of the injured as dictated by the nature and severity of the injury.

Once care is assured, notification of the third supervisory level is timely. With today's high-tech communications capabilities reporting can be by cell-phone or radio. The object is that the injury notification reaches the top management within hours, if not minutes, after the injury.

3. Investigating the injury.
As soon as the injured party receives appropriate medical care the investigation of the accident leading to the injury begins immediately. This means no later than the day of the

injury even if the injury occurs late in the workday. Injury to an employee justifies those involved in the investigation to work overtime if need be to get the process started. Delay in investigation allows needed details to slip away.

Every employer should have a pre-selected accident/injury investigation team always on alert in case there is a need. These personnel know in advance they are responsible to begin the investigation immediately. In a manufacturing facility the team members are appropriate leaders and safety personnel. Any technical support should be added as required. Line management should lead this investigation. This is part of the accountability aspects of safety in the workplace. Even if the injured employee's foreman is still tied up with tending to the injured the investigation can begin.

ALERT: This is a fact finding investigation aimed at understanding cause and effect, NOT a session to determine who is deserving of some kind of discipline for violating some safety code. Any contemplated punishment should take a back seat to the getting the facts. An injury investigation team should not be the group who determines appropriate disciplinary action. This aspect of an accident belongs to line management, not the investigating team.

First priority is to make a list of witnesses and co-workers in the immediate area. Then an investigation strategy is developed; first things first etc.

Too many times such responsibilities are "handed off" to the safety department. This is not appropriate. Management has the most to learn with the primary objective being to prevent a similar accident or injury due to the lack of swift and decisive management intervention. However, safety must always be represented on the team.

4. Tending the injury and the injured.

If the injury requires a medical professional or physician there are four management considerations that need attention. These are: a. Pre-selected care-givers and facilities, b. Employee communication as to who and where these facilities are located with phone numbers and maps, c. Responsible company personnel in attendance at the facility interfacing with the care-givers, d. Post injury management.

 a. It is always a good idea to pre-select from the local area medical professionals those you will use in caring for any injured employees. Interviews with these professionals outlining your priorities: of course the first of these is that your employee receive the finest care available. The second is to assure the professional that you are not trying to intervene in the appropriate care of the injured employee. That you are there for counsel on the nature of the employee's work, work-place and any alternate work that may be available in case the injured is placed on limited duty. That you also are here to assist the injured's family members in tending to their loved one.

 b. Employee communication as to who and where these facilities are located with phone numbers and maps.
All supervision should have printed information, phone numbers and maps on the location of care facilities and medical personnel that are pre-selected.

 c. Responsible company personnel in attendance at the facility interfacing with the care-givers. This can be the foreman or other supervisory personnel. It also can be the safety person in some situations, however to delegate this important activity to the safety person dilutes the line management accountability, so it is not recommended as the norm.

 d. Post injury management
If an injury is severe enough to require time away from work there should be a coordinated company

259

wide process that continues to assure the injured receives proper medical care as well as routinely remains in contact with the injured in a care-giving mode. Such a mode is important to impart to the employee the sincerity of the employer regarding the injured employee's well-being and recuperation. As appropriate return to work possibilities are also discussed. Return to work is determined by the medical care-giver's advice to the injured and the employee's department as to available limited duty work opportunities, if needed.

Part of the post injury management process is to be actively engaged with the Workers' Compensation insurance carrier to ensure the case is managed to your standards and your areas of concern for the care of your employee is given attention. It is always a good practice to be involved with the carrier's loss manager to discuss appropriate reserves for injuries requiring extended care.

9. Drug and alcohol testing

Most enterprises today utilize some sort of Drug Testing Program. Some of these are very good and some, for a number of reasons, do not go far enough to ensure a drug free working environment. There are three desired features of a Drug and Alcohol Testing Program.

These are;
1. Pre-Employment Testing,
2. Random Testing, and
3. Post Accident Testing.

1. Drug Testing: A well thought out Drug Testing policy should be a part of every company that embraces the

Zero Injury Concept. During the past decade the Drug Testing industry has expanded to where there are a number of qualified laboratories to handle the testing of specimens.

Pre-employment screening is mandatory in my view. On saying this I know a few companies who by policy do not allow their employees to be tested for reasons of personal privacy. This of course is their right. Utilizing such a company may give rise to other policy considerations by Owners or General Contractors.

The subject gets complex with legal implications so I will not offer comment other than to say your policy is your policy and their policy is their policy. If your policy says testing is required you do not have to employ a company that is not of like mind and purpose. There are likely exceptions again where municipal work is the object. Be aware that some local governments by statute do not allow drug testing even on private work within jurisdiction.

2. Random Testing: It is my belief that random drug testing is a vital part of a drug policy. While there are many examples of testing policies where random testing is not done, the best of policies do include such testing. The object of course is to discourage as much drug use as feasibly possible.

To not random test is to know that on any given day some numbers of your employees are sufficiently affected by the previous evening's drug use to be a danger to themselves and those that work around them. There is reluctance on the part of some Unions to allow random testing of represented employees. This is argued as a personal right of the employee to be free from invasive inspections into their personal habits.

On the other hand some Unions operate their own testing programs with some having random testing as a part of the Union's own policy. In other instances, particularly in the construction industry, where there is a policy of no random testing that this provision is waived by the Union when the Owner of the project requires random testing of all contractor employees.

3. Post-Accident Screening: Post accident drug testing is a productive policy feature. To not drug test after an accident is to omit an obligation to your employees who are drug free. Their safety is your immediate concern and often accidents do occur as a result of drug use. There is only one way to know: test.

CLOSING THOUGHTS

The Zero Injury discovery by the Business Roundtable Construction Committee in 1987 has been perhaps the most significant advance in safety program leadership in the past century. The moral concept of conducting an enterprise without injury to employees has caught fire literally around the world. In truth no one in 1987 would have dreamed of the amazing safety advances over the past 20 years growing out of the Zero Injury Safety Leadership Concept.

The Zero Injury Institute (ZII) reports that since 1989 in the USA they have learned of sixteen contractors that have exceeded 1,000,000 hours worked in a string with Zero Recordable Injuries. ZII has placed their names on the ZII Safety Hall of Honor for this world class safety performance. ZII recognizes there surely are others they do not have in their records and asks any that wish to be added to contact ZII via the ZII web site www.zeroinjuryinstitute.com.

These remarkable few are listed below as of Feb. 22, 2011.
1. Zachry, 1,020,000, Shell Chemical, LA,1989
2. Fluor, 2,080,000, CITGO, Lake Charles, LA, 1996
3. NPS Energy Services*, 1,000,000, Peach Bottom Power Plant, York County, PA,1999
4. Cherne Contracting*, 2,547,248, Minneapolis, MN, Company Wide, 2000-2001
5. Parsons, 1,229,585, Exxon, Baytown, TX, 2002
6. J. H. Kelly*, 1,036,746, Conoco Phillips, OK, 2005
7. S&B E&C, 4,649,000, Houston, TX, Company Wide, 2003-2004
8. Burns-McDonnell, 1,000,000, Conoco Phillips, Borger, TX, 2005
9. Cherne Contracting*, 1,500,000. Sunoco & Ky Power, 2005
10. Superior Construction*, 1,000,000, BP Whiting, IN, 2006
11. Solid Platforms, Inc.*, 2,569,257, BP Whiting, IN, 2007
12. McCarl's Inc.*, 1,020,785, York Haven, PA, 2008
13. Chapman Corporation*, 1,081,486, Canonsburg, PA, 2009

14. D. E. Harvey, 1,149,142, San Antonio, TX, 2009-2010
15. Cherne Contracting*, 1,700,000, Minneapolis, MN, 2011 ongoing.
16. Kiewit-Southwest*, 3,600,000, Phoenix, AZ, 2011 ongoing

   * Union

I here record my sincere congratulations to these companies and the multiple owners for whom they performed work for their stalwart safety achievements. As a result of their singular efforts and that of hundreds of other contractors and owners we can all realize that hundreds of injuries have been averted and perhaps fatalities have been avoided as a consequence of their safety commitment! When we attempt to get an exact accounting of these fortunate unharmed workers the only conclusion one can reach is expressed by the following observation.

> We will never know how many lives have been saved and how many millions of injuries averted by this one significant safety advance from the idea of one man, Ed Donnelly of Air Products who advanced the simple notion that in one industry safety could be improved by the feature of recognizing those companies who were accomplishing safety excellence. By this recognition it became known for the first time that safety excellence was re-defined in 1987 as "zero injury!"

Emmitt J. Nelson,
March 2011
Houston, Texas
nelsonci@att.net